Springer Geography

The Springer Geography series seeks to publish a broad portfolio of scientific books, aiming at researchers, students, and everyone interested in geographical research.

The series includes peer-reviewed monographs, edited volumes, textbooks, and conference proceedings. It covers the major topics in geography and geographical sciences including, but not limited to; Economic Geography, Landscape and Urban Planning, Urban Geography, Physical Geography and Environmental Geography.

Springer Geography — now indexed in Scopus

Vishwambhar Prasad Sati

Exploring the Mystical Himalaya

Cultures, Economies and Environment

Springer

Vishwambhar Prasad Sati
Department of Geography and Resource Management
Mizoram University
Aizawl, Mizoram, India

ISSN 2194-315X ISSN 2194-3168 (electronic)
Springer Geography
ISBN 978-3-032-14274-0 ISBN 978-3-032-14275-7 (eBook)
https://doi.org/10.1007/978-3-032-14275-7

This Springer imprint is published by the registered company Springer Nature Switzerland AG
The registered company address is: Gewerbestrasse 11, 6330 Cham, Switzerland

Preface

During my three decades of academic career, I have composed more than two dozen treatises on different aspects of the Himalaya. These works span various publishers, including Springer Nature, Elsevier, Bishen Singh Mahendra Pal Singh, Today and Tomorrow Printers and Publishers, Mittal Publications, and Cambridge Scholars Publishing. My publications cover diverse themes such as sustainable agriculture, natural and cultural diversity, forest management, environmental significance of the Ganges, tourism development, mountain ecosystems, resource management, society and culture, pilgrimages, horticultural development, and sacred temples. The books explore comprehensive perspectives on the Himalayan region, examining its ecological, social, economic, and cultural dimensions. Complementing these book-length works, I have also published over 160 impactful research papers that further contribute to the academic understanding of the Himalaya, reflecting a sustained and significant scholarly engagement with this complex and fascinating geographical region.

The Himalaya holds rich cultures, economies, and environment. It is not only a natural entity but also a cultural destination. Its importance is presented in various Hindu scriptures. In the Shrimad Bhagwat Gita, while preaching to Arjuna and describing his various forms, Lord Krishna says:

> महर्षीणां भृगुरहं गिरामस्म्येकमक्षरम् ।
> यज्ञानां जपयज्ञोऽस्मि स्थावराणां हिमालयः ॥ (२५)
> मैं महर्षियों में भृगु हूँ। मैं सभी वाणी में एक अक्षर (प्रणव) ॐ हूँ। मैं सभी प्रकार के यज्ञों में जप (कीर्तन) यज्ञ हूँ, और मैं ही सभी स्थिर (अचल) रहने वालों में हिमालय पर्वत हूँ। (२५)
>
> I am Bhrigu among the great sages. Among all forms of speech, I am the single syllable (Pranava) "Om." Among all types of sacrifices, I am the sacrifice of chanting (Japa Yajna). And among all immovable things, I am the Himalaya. (10.25)

The Himalaya is the source of three of the world's largest river systems—the Sindhu (Indus), the Ganges, and the Brahmaputra. These river systems are the lifelines of South Asia, supporting more than 600 million people. They hold natural, cultural, economic, and environmental significance. The Ganga River, in particular, holds deep spiritual importance. Hindus regard it as a mother and call it "Mother

Ganga". Both the Himalaya and the Ganga River are culturally intertwined and are frequently mentioned together in many Hindu scriptures. The scriptures say:

> हमिाद्रेः सम्भूता वहिरसतिदके हमिमये,ततो भूमयिाता वचिरससिुहासा समतले। पुनासत्विं लोकान् प्रवहससिमुद्रं प्रतसिदा, अहो दवि्ये गङ्गे! भवतु सुखदं नस्तव जलम् ।।
>
> हे गंगा माता! तुम हमिालय से उत्पन्न हुई, हमिालय की गोद में वचिरण करती हो; तब पृथ्वी पर आती हो और समतल सन्न होकर हँसते हुए वचिरण करती हो।
>
> O Mother Ganga! You originate from the Himalaya, wander in the lap of the Himalaya; then you descend to the earth and flow gracefully over the plains, meandering joyfully.

The treatise *Exploring the Mystical Himalaya: Cultures, Economies and Environment* is a comprehensive exploration of the Himalayan region, documenting my extensive travels from the Pamir's knot to the Arakan Yoma, including Tibet, Bhutan, and Nepal. Based on personal observations and travels through all Indian Himalayan states, Nepal, Bhutan, and Trans-Himalayan regions like Ladakh and Tibet, the book provides an in-depth narrative focusing particularly on Uttarakhand. The nine chapters cover a range of themes from the mystical nature of the Himalaya to specific regional experiences, exploring cultural, geographical, and spiritual dimensions through travelogues that highlight the unity, diversity, and interconnectedness of this magnificent mountain range, concluding with reflections on borders, cultures, and the profound essence of the Himalayan landscape.

The Himalaya is a complex, multifaceted region encompassing rich natural, economic, social, and cultural dimensions. Beyond its geographical significance, it is a repository of diverse cultures, customs, rituals, festivals, human races, and an extraordinary array of flora and fauna. The mountain range hosts thousands of peaks, rivers, basins, streams, and trails. Critically, the Himalaya sustains human life across upstream and downstream valleys, preserving cultures, customs, and human races since prehistoric times. It serves not just as a natural landscape but as a vital ecological and cultural guardian, protecting the country and supporting human civilization through its intricate and interconnected ecosystems.

This book is based on my travelogues about different parts of the Himalaya and Trans-Himalaya. In 2016, I travelled to Lhasa and other parts of Tibet twice and to China several times, studying various natural and cultural aspects of Tibet and Tibetans, respectively. In 2022, I travelled to Leh and Ladakh, visiting Leh, Nubra, Pangong, and Kargil, and composed a detailed travelogue. In 2023, I journeyed to Nepal and travelled from Kathmandu to Pokhara and beyond along the Kali Gandaki, a holy river of Nepal. In 2024, I made a wonderful visit to Bhutan, traveling from Paro International Airport to Thimphu and visiting many cultural and natural destinations. Uttarakhand (the Central Himalaya) is my home state, where I have travelled thousands of times to numerous places, with only the Kali Kumaon part left unexplored. I divided the Uttarakhand Himalaya into three parts—Hari Ki Doon (the Valley of God), the Kedarkhand (further divided into three parts), and the Manaskhand. I have travelled through the trails of Kedarkhand many times, while Hari Ki Dun was trekked in July 2024. It was a seven-day trek from Dehradun to Mussoorie, Naugaon, Purola, Mori, Hanol, Netwad, Sankari, Dharkot, Gangad, Osla, and Hari Ki Dun.

In the last week of December 2022, I explored three districts of Kumaon region—Almora, Bageshwar, and Nainital. I journeyed through the Sun Temple, Kasar Devi Temple, Jageshwar group of temples, Mukteshwar, and seven lakes of Nainital district. Similarly, in December 2024, I travelled to the Johar, Darma, and Vyas valleys of the Kumaon Himalaya. I started my journey from Dehradun and passed through Rishikesh, Devprayag, Srinagar, Karnprayag, Narayan Bagar, Gwaldam, Baijnath, Bageshwar, Kapkot, and Munsiyari (Johar Valley). After that, the journey continued to Joljivi, Dharchula, and the Darma Valley, with the last destination being Dartu and the Panchachuli base camp and zero point. From Tawa Ghat, another route goes to the Vyas Valley, renowned for the Gunji and Kuti villages, and the spiritual shrines of Adi Kailash and Om Parvat. This was the last trail, and with this, I covered all trails of the Himalaya and finalized the book.

While reflecting on the cultural, economic, and environmental significance of the Himalaya and staying in Dartu village, Darma Valley (3200 m altitude), Kali Kumaon (26 Dec 2024), I composed these lines:

O Himalaya, majestic and grand,
With countless peaks, you grace the land.
Your glaciers feed rivers that endlessly flow,
Life's nourishment in your heart does grow.
Your rivers weave tales of sustenance and care,
As waterfalls whisper joy in the air.
Your moods shift in hues and celestial light,
A canvas of wonder, from dawn to night.
Cultures and customs, you cradle with grace,
Sheltering human hearts in your vast embrace.
Fauna and flora thrive in your fold,
A sanctuary of life, both tender and bold.
O Himalaya, guardian of skies so high,
Your spirit eternal, no soul can deny.

This book is valuable to all stakeholders—students, scholars, academicians, teachers, pilgrims, and tourists. It will help readers understand the Himalaya and its natural and cultural landscapes, offering insightful knowledge through my travelogues. Finally, I dedicate this book to my parents and family members.

Aizawl, Mizoram, India
January 1, 2025

Vishwambhar Prasad Sati

Competing Interests The author has no competing interests to declare that are relevant to the content of this manuscript.

About the Book

The treatise *Exploring the Mystical Himalaya: Cultures, Economies and Environment* is composed based on the collection of empirical data, extensive field visits, and thereafter my personal observation through travelogue of the entire Himalaya from Pamir's knot to the Arakan Yoma (including Tibet, Bhutan, and Nepal) with aspects of nature and culture. It presents a comprehensive illustration of the Himalaya. There are nine chapters—the Introduction: the Mystical Himalaya; Ladakh: The Land of Lamas; Hari Ki Dun: The Valley of Lord Vishnu; The Legends of Kedarkhand: A Journey Through the Himalayan Abode; Manaskhand: The Climax of Nature and Culture; Kathmandu to Muktinath: Unity and Continuity Through the Kali Gandaki River; Harmony in the Himalaya: Buddhism and Bhutan; Beyond Borders and Culture; and Conclusions. Various aspects of the Himalaya—natural, economic, social, cultural, and finally the customs and rituals performed during the off-time—are illustrated in all chapters. I observed that the Himalaya is not only a natural entity but it also holds high cultural significance. It is the land of a thousand cultures, customs, rituals, fairs, festivals, human races, and faunal and floral species. Furthermore, it holds thousands of mountain peaks, rivers, river basins, streams, rivulets, and trails. The Himalaya preserves mankind through supporting lives of thousands of people in upstream and downstream river valleys. It has been preserving the culture, customs, rituals, and human races since prehistoric times. It is a warier, protector from the neighboring countries. The book is valuable for all stakeholders—students, researchers, academicians, policy makers, pilgrims/tourists, and the Himalayan lovers.

Contents

About the Author

Vishwambhar Prasad Sati born in 1966, holds a Doctor of Letters (D.Litt.) and Ph.D. degrees, currently serving as a Senior Professor of Geography and Resource Management and Dean of Faculty, School of Earth Sciences and Natural Resources Management at Mizoram University, a Central University in Aizawl, India. With more than three decades of teaching experience, he has contributed to various higher education institutions within and outside India, including the Eritrea Institute of Technology in Asmara, Eritrea (NE Africa); the Institute of Mountain Hazards and Environment in Chengdu, China; and HNB Garhwal University in Srinagar Garhwal. Throughout his illustrious career, he has actively engaged in research, holding prestigious positions as a Presidential (PIFI) Fellow/Visiting Professor (2016) "Chinese Academy of Sciences"; Visiting Scientist (2014) "Chinese Academy of Sciences"; Visiting Scholar (2010) "The World Academy of Sciences" (TWAS); Visiting Scientist (2012) "Indian National Science Academy" (INSA); General Fellow (2008-2009) "Indian Council of Social Science Research" (ICSSR); and Associate (2008) "Indian Institute of Advanced Study" (IIAS). He is recognized as one of the 0.5% researchers of the world and 50 top influential academicians of India. Furthermore, he has received recognition for his contributions, including a Visiting Professor Award from the Ministry of Education and Science at Lviv Polytechnic National University in Lviv, Ukraine; IMPRI India; a Best Paper Award from Natural Hazards Research (an Elsevier Journal); Best Presentation

Award in the International Conference, held at Jurich, Switzerland; and the title of Senior Global Fellow from the Grassroots Institute in Canada. His extensive academic travels, spanning over 50 countries and all over India, have contributed to his comprehensive understanding of diverse geographical landscapes. Additionally, he has successfully supervised 14 Ph.D. scholars. His expertise lies in Natural Resource Management, Environment-Development Interface, Climate Change, Rural Livelihoods, Disaster-Related issues, and the Culture of the Himalayan region. Vishwambhar Prasad Sati has developed numerous models and action plans for environmental conservation and economic development in the fragile Himalayan ecosystems. His impactful contributions extend to more than 200 research papers and articles and over three dozen published books, establishing his academic legacy at both national and international levels.

List of Figures

Chapter 1
Introduction

Contents

Abstract The Himalaya stretches from the Pamir Knot to the Arakan Yoma mountain ranges of Myanmar. It is the tallest and youngest mountain range in the world. The region is rich in both natural and cultural diversities, which varies vertically and horizontally across its vast expanse. This study was conducted using case studies and the collection of empirical data. The author visited the entire Himalayan region—including Leh-Ladakh, Nepal, Bhutan, and Tibet—and composed this book based on extensive observations. This chapter provides a brief overview of the landscape, climate, forests, culture, society, and people of the Himalayan region. It also outlines the objectives of the study, the methodology employed, and the organization of the book.

Keywords Landscape · Climate · Natural vegetation · Society · Culture · Himalaya

V. P. Sati, *Exploring the Mystical Himalaya*, Springer Geography,
https://doi.org/10.1007/978-3-032-14275-7_1

The Himalaya Mountain stretches over 2500 km from Pamir's Knot (northwest) to the Arakan Yoma Mountain Ranges of Myanmar (southeast), covering the Jammu and Kashmir Himalaya, Himachal Himalaya, Uttarakhand Himalaya, Nepal Himalaya, Sikkim Himalaya, Bhutan Himalaya, and Arunachal Himalaya (Aitken 2003). Ladakh and Tibet are part of the Trans-Himalaya. The landscape of the Himalaya is varied and spectacular (Sati 2019). It has different colors, shades, moods, climates, landforms, forest types, human races, cultures, customs, and rituals. All these make the Himalaya a rich and unique entity (Sati 2024). Snow-capped mountain peaks, huge glaciers, alpine grasslands, high-altitude deserts, spectacular river valleys, rich biodiversity, rich agrobiodiversity, and rich cultural heritage further enhance the beauty of the Himalaya (Sati 2022).

The Himalaya has both natural and cultural significance (Sati 2023; Swami 1960). On the one hand, it is a source of three major river systems in the world, including the Sindhu system, the Ganges system, and the Brahmaputra system with their numerous tributaries (Sati 2021). On the other hand, it is home to a rich cultural heritage. The three river systems of the Himalaya not only provide ample water for drinking and irrigation but also support the livelihoods of over 600 million people living in upstream and downstream areas. The Himalaya is renowned as the water tower of the world and the third pole. It is one of the world's 12 highest biodiversity hotspots. In addition, it is believed to be the abode of Hindu deities—Shiva and Shakti, and the land of Buddha. Ladakh, Tibet, and Bhutan are the lands of the Lamas. Meanwhile, the Indian Himalayan region, mainly the Central Himalaya, is believed to be the land of Shiva and Shakti. Nepal follows both Buddhism and Hinduism; therefore, the folk deities are related to both religions. The two states of the Eastern Himalaya—Sikkim and Arunachal—follow both Hinduism and Buddhism. A small portion of the population follows Sikhism, Jainism, Islam, and Christianity.

In this chapter, a detailed illustration of the natural and cultural aspects of the Himalaya and Trans-Himalaya, including Uttarakhand, Nepal, Bhutan, Ladakh, and Tibet, is carried out. This study is based on my travelogs and experiences gained from my extensive visits to almost the entire Himalayan region. This study includes landscapes, landforms, mountain peaks, rivers, glaciers, forests, national parks and wildlife sanctuaries, agriculture, livestock, tourism, culture, customs, rituals, fairs and festivals, folk deities, and the mythology of worshipping folk deities.

1.1 Landscape and Landforms

The Himalaya includes the river valleys, the Lesser Himalaya, Middle Himalaya, Greater Himalaya, and the Trans-Himalaya vertically. It has 14 peaks higher than 8000 m and 200 peaks over 7000 m (Sale and Cleare 2000), some of which are Everest, Kanchenjunga, Nanda Devi, Kamet, Trishul, Chaukhamba, Panchachuli, and Annapurna (Fig. 1.1). Everet is the highest peak of the world, situated in north part of Nepal, border with Tibet. The Uttarakhand Himalaya has many popular

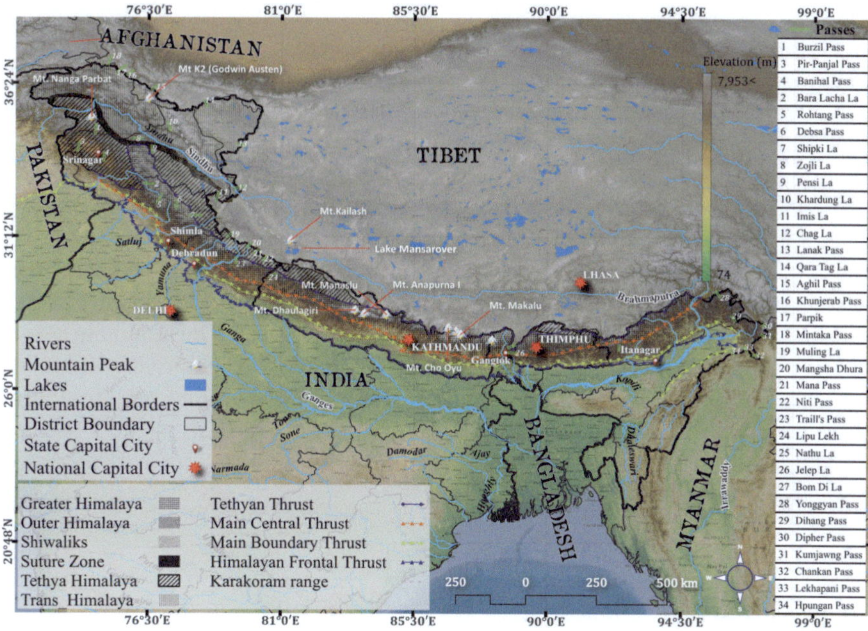

Fig. 1.1 The mystical Himalaya: different landscapes, river systems, mountain peaks, and trails

mountain peaks, which include Nanda Devi, Kamet, Trishul, and Chaukhamba. In Nepal Panchachuli and Annapurna are main mountain peaks. Kanchenjunga is the famous mountain peak of the Eastern Himalaya located in Sikkim border with Nepal. Horizontally, the Himalaya has three divisions—the Western Himalaya, the Central Himalaya, and the Eastern Himalaya. The Western Himalaya comprises of Jammu and Kashmir, Himachal, and Uttarakhand. Nepal is in the center, and Sikkim, Bhutan, and Arunachal are located in the Eastern Himalaya. Tibet and Ladakh are located in Trans-Himalaya, the highest tablelands and cold deserts of the world.

The Himalaya is manifested in its landscapes, rivers/streams, snow-capped mountain peaks, and in rich forest diversity (Joshi et al. 2025). The spectacular landforms; humming rivers/streams, making different river topographies cascades, waterfall, rapids, and meanders; calling mountain peaks with changing their colors and forms; and a panoramic and diversified forest landscapes, making different forest layers; and chirping of animals and birds are the important features of the Himalaya (Valdiya 2016). In addition, the Himalaya is the taposthali (center for penance) for sages and saints, heaven for pilgrims, and supporting life to billions of people. The Kali Gandaki makes the deepest gorge in its source area, middle Mustang, and the Jahanvi River makes a longest gorge at Bhaironghati before meeting with the Bhagirathi River. The Alaknanda River makes a narrow gorge at Koteshwar near Rudraprayag town. All Himalayan rivers make beautiful meanders. The Ganga River makes a longest meander at Tota Ghati near Devprayag and at the confluence of the Nayar River and the Ganga River. Near the source area, all Himalayan rivers make waterfalls. One of the biggest waterfalls is Vasudhara, located in the Saraswati River, on the way to Swargarohini Mountain (Rao et al. 2008). The Bhagirathi River makes beautiful cascades near Gangotri. "U"-shaped valley in the snow area and "V"-shaped valley in the river valleys are beautiful landscapes in the Himalaya.

Stretch valley of Brahmaputra, river terraces of the Alaknanda River, and potholes in the source area of the Bhagirathi Rivers are the other landforms of river valleys. The Trans-Himalayan region holds different landscapes than the Himalayan region. The Leh-Ladakh region of Indian Himalaya and Upper Mustang region of the Nepal Himalaya are cold deserts and they form sand dunes. The mountain ranges are baren and stony, on the tope of which are snow-capped. The river valleys are deep with high water volume and velocity. Oasis (green grassland with some tree species) is found in the areas where water is available. These rivers flow through deep gorges and wide river valley plains. Similarly, these rivers make many waterfalls and cascades.

1.2 Himalayan Climate and Its Implications

The Himalayan climate is diverse and varies significantly with altitude. Climatic zones range from subtropical to sub-temperate, temperate, cold, and frigid, with considerable seasonal variability (Wester et al. 2019). The Middle Himalaya, stretching across the Himalayan region, remains pleasant primarily during the summer season but becomes cold in winter. The alpine meadows and the High Himalaya are snow-covered for six months during winter, while the Greater Himalaya remains perpetually snow-capped. The river valleys experience hot and dry conditions during the summer, with winters being generally mild, although occasional cold waves occur. Rainfall predominantly happens during the monsoon season but is also observed in winter, known as western disturbances. Humidity remains high throughout the year, particularly in the highland areas. The Himalayan climate significantly impacts the Gangetic plains and Central India, influencing both summer and winter conditions (Parashar et al. 2024). Heavy snowfall in the Himalaya during winter often triggers cold waves in downstream regions. Recently, climate change effects have become noticeable in the Himalaya. Valley regions and middle altitudes are warming, and rainfall intensity and frequency have increased, though its duration has shortened. These changes have led to more frequent and severe geohydrological hazards (Belho et al. 2024). Additionally, agricultural systems and biodiversity have undergone notable transformations in recent years.

1.3 Himalayan Glaciers and River Systems

Himalaya holds a large snow cover and many glaciers. These glaciers are the sources of the major river systems in the world (Kulkarni and Karyakarte 2014). Two Himalayan rivers—Sindhu and Brahmaputra originate from two glacial fed lakes—Rakshas and Mansarover, respectively. Bandarpunch, source of the Tons River; Yamunotri, source of the Yamuna River; Gaumukh, source of the Bhagirathi River; Chaurabari, source of the Mandakini River; Sundardhunga, source of the Nandakini River; Pindari, source of the Pindar River; Namik, source of the Ramganga (E); Zemu, source of the Teesta River, and many other glaciers make the Himalaya the water tower and the third pole (Oli and Pandey 2024). The rivers of Nepal are also glacier-fed. During the summer and rainy seasons, these rivers overflow often above the danger marks (Shrestha et al. 2015).

1.4 Flora, Fauna, and Avifauna

The forests of the Himalaya are unique and highly diversified. Altitude, followed by climate, determines the types and distribution of forests along the altitudinal gradient. From the valleys to the high altitudes, forest types vary significantly (Negi et al. 2024). In the river valleys, below 800 m in altitude, subtropical deciduous forests, primarily consisting of teak varieties, are found. Between 800 m and 1600 m (in the middle Himalaya), pine forests dominate, covering approximately 28% of the total geographical area. The third forest zone is located in the highlands between 1600 m and 2200 m, where mixed-oak trees thrive. Temperate coniferous forests are found between 2200 m and 2800 m, marking the last tree line, beyond which alpine grasslands spread above 2800 m. The final landscape comprises snow-capped mountains, starting from around 3200 m. The entire Himalayan region is rich in biodiversity and is recognized as one of the 12 biodiversity hotspots in the world. In the Trans-Himalayan region, forest types and coverage are minimal, found only in oasis areas (Nandy et al. 2006). The Himalaya also supports a rich faunal diversity, including a wide variety of avifauna. Among the fauna, species such as elephants, peacocks, tigers, panthers, wild boars, chitals, reindeer, snow leopards, wild pigs, and many others dominate at different altitudes and latitudes. These elements together create a panoramic forest landscape.

1.5 National Parks and Wildlife Sanctuaries

The Himalaya holds many national parks and wildlife sanctuaries, spanning from northwest to southeast. Nanda Devi Wildlife Sanctuary, Nanda Devi National Park, Govind Wildlife Sanctuary, Rajaji National Park, Corbett National Park, and many more are located in the Central Himalaya. The Kangchenjunga Conservation Area covers 2035 km^2. Mount Kangchenjunga (8586 m), the world's third-highest mountain, is renowned for its alpine meadows, high-altitude wetlands, glaciers, and biodiversity, including red pandas and snow leopards (WWF 2024). Jigme Dorji National Park of Bhutan is also home to many mountain peaks, which are considered sacred (Devi et al. 2021). This is a protected area. The high-altitude wetland areas in India are considered sacred places. The mountain peaks, forests, and rivers are also considered sacred. Many fairs and festivals are celebrated in these sites every year. The strong bond between nature and culture plays a significant role in habitat conservation. In Sikkim and Western Arunachal Pradesh, Buddhist monasteries protect sacred groves. In a nutshell, the followers of Hinduism, Sikhism, Jainism, and Buddhism are conservators of nature.

1.6 Hydropower Projects

The Himalayan region is home to renewable energy sources such as micro-hydro power projects, solar, and wind energy. The Terai Arc of the Himalayan region has high potential for producing biogas. The vast reservoir of snow, numerous glaciers, and glacier-fed rivers provide significant bases for the construction of hydroelectric power projects (Aryal et al. 2024; Desai et al. 2023). There are many world-famous and tall hydropower projects, among which the Tehri High Dam is prominent and the second tallest in Asia. The major dam projects are constructed on the Sindhu, Sutlej, Yamuna, Ganga, Gandaki, Tista, and Brahmaputra Rivers.

There are more than 200 hydropower projects—constructed, under construction, and proposed in the Central Himalayan Region—which produce more than 10,000 MW of hydroelectricity (Sati 2014). The Himalaya supplies hydropower energy abundantly to other parts of the country. The region also has many ideal locations for generating solar and wind energy. Meanwhile, the entire Himalayan region is ecologically fragile and highly prone to natural hazards. Therefore, the construction of large-scale hydropower projects is not suitable for the Himalayan region. There have been many instances of significant destruction caused by the collapse of hydropower projects, such as the Vishnu Gad Tapovan calamity of 2021, which killed more than 200 laborers who were working on dam construction (Sati 2021). It is recommended that micro-level dam projects be constructed in the Himalayan region to avoid any future catastrophes.

1.7 Social Systems

The Himalayan region has rich linguistic diversity, with the principal language families being Indo-European and Tibeto-Burman (Choudhary and Garkoti 2023). Meanwhile, there are many languages spoken across the entire Himalayan region. A proverb aptly states, "At every mile, water changes, and at every four miles, language." Other prominent languages include Garhwali, Kumaoni, and Jaunsari. In addition, people from each drainage basin speak their unique languages. The Greater Himalayan region is predominantly inhabited by Tibetans, who speak Tibeto Burman languages, whereas the Lesser Himalaya is home to speakers of Indo-European languages. The Kashmiri people of the Vale of Kashmir and the Gaddi and Gujari communities of the Lesser Himalaya is hill people who rear goats and sheep and practice seasonal migration (Mishra et al. 2023). During the summer, they migrate to high-altitude pastures, and when the high-altitude areas of the Outer Himalaya receives heavy snowfall, they move to the river valleys with their animals.

Gujari, Champa, Ladakhi, Balti, and Dard people live in the northern part of the Greater Himalaya (Upper Indus Valley). They are traditionally nomadic pastoral communities. Bhotiyas, Jadh, Lepcha, Pahari, Tamang, Kanet, Khasi, Gurung, and Sherpa are other mountain communities who live in the high Himalayan region, rear livestock, and practice shifting cultivation. The Gaddis of Himachal Pradesh, the Lepchas of Sikkim, and the Bhotiyas of Uttarakhand are transhumance communities. They encompass diverse ethnic groups with distinct and varied cultures, traditions, and beliefs (Kandari and Gusain 2001).

The social structure comprises Brahmins, Rajputs, Scheduled Tribes (ST), and Scheduled Castes (SC) (Sati 2023). Brahmins and Rajputs have a higher population and are considered the socially privileged class. Brahmins perform *puja* on auspicious occasions and practice agriculture. Rajputs generally serve in the National Army. They also own agricultural fields and rear livestock. STs and SCs are indigenous and belong to the unprivileged class. However, with social reforms introduced by the governments of the Himalayan region, the social and economic conditions of STs and SCs have improved. They possess arable land where they practice agriculture. However, the STs mainly focus on livestock farming, and many of them are known as nomads. They maintain seasonal dwellings in the highlands and river valleys. During the summer, they move to the highlands with their animals—mainly goats, sheep, lambs, horses, and a few cattle—to graze them in alpine pasturelands. The climate remains pleasant during the summer as the snow melts, allowing a variety of grasses and flowers to grow, which are highly suitable for rearing animals (Singh and Kerven 2023). Meanwhile, during the winter season, the high-altitude regions receive heavy snowfall, forcing the animal herders to move down to the valley regions. They also practice shifting cultivation. This process is referred to as highland-lowland interaction and seasonal migration. The SCs traditionally worked as landless laborers in the agricultural fields of the privileged class. They played drums during fairs, festivals, and auspicious occasions, collecting food grains from the privileged people in return. However, over time, SCs have acquired agricultural land and now also work in the tertiary sector. The nomads of Ladakh, the highland areas of Uttarakhand, the Upper Mustang Valley of Nepal, and the highlands of Bhutan, also known as transhumance, migrate seasonally (Guha 2000). They move from river valleys to highlands along with their cattle.

1.8 Economy and Development

The farming community of this region is marginal and practices agriculture in the river valleys, middle altitudes, and highlands on narrow and fragmented arable lands. Farming is also practiced on the terraced lands of the Lesser Himalaya and Foothills. The economy of the Himalayan people is primarily dependent on traditional agriculture and livestock rearing (ICIMOD 2024). However, the arable land in the Himalayan region is very limited, ranging from 10% to 15%. A large portion of the arable land is rainfed, with approximately 10% being irrigated. Moreover, the proportion of rainfed and irrigated land varies between the plains, river valleys, middle altitudes, and highlands. In the plains and river valleys, a significant portion of the arable land is irrigated. Conversely, the middle and highland areas are predominantly rainfed.

Agrobiodiversity in the Himalaya is rich, encompassing more than 50 species of food grains, pulses, oilseeds, fruits, and vegetables. The crop diversity varies significantly from the river valleys to the middle and higher altitudes (Bijalwan et al. 2018). While crop diversity is relatively low in the river valleys, it is notably high in the middle and high altitudes. Paddy and wheat are primarily grown in the river valleys, whereas varieties of fruits, vegetables, oilseeds, and pulses are extensively cultivated in the middle and highlands. Meanwhile, the production and productivity of crops are quite low, which is insufficient to adequately feed the rural population. Consequently, the Himalayan region faces food scarcity and malnutrition. The reasons for low production and productivity include rough and rugged terrain, rainfed and terraced agriculture, high slope erosion, and limited opportunities to apply modern innovations in agriculture.

The Himalaya has abundant forest resources, and consequently, the non-timber forest products (NTFPs) are plentiful, forming the basis of livelihoods for rural people living in the remote Himalayan region (Sherpa 2025). Timber is widely used for constructing houses, making furniture, and as firewood. Grasses from the subtropical and temperate grasslands are utilized for feeding animals. Furthermore, leaves from oak and other trees are highly useful for feeding livestock and enhancing milk production. The bark of oak trees is used to produce spices and ammunition. Tejpatta (Cinnamomum tamala) is a valuable spice found in the Himalayan forests. A variety of wild fruits are available in the temperate and subtropical forests, which possess high nutritional value. Medicinal plants are also abundant in the temperate regions and alpine grasslands. All these NTFPs significantly support the livelihoods of the native population.

Tourism has several types practiced in the Himalayan region. Cultural tourism (pilgrimage), natural tourism, adventure tourism, and eco-tourism are prominent. It plays a greater role in the economic development and livelihood enhancement. Mainly pilgrimages to the Himalaya are the centuries-old practices. The Hindus and Buddhists throughout the Himalayan region have highland temples and monasteries where they visit to pray Gods. Kailash-Mansarover in Tibet, Char Dhams of the Uttarakhand Himalaya, and Muktinath in the Upper Mustang Valley in the Nepal

Himalaya are the prominent highland pilgrimages for the Hindus (Chand and Mishra 2024). Similarly, in Ladakh and Bhutan, monasteries are the sacred places of worshiping Lord Buddha for the Buddhists. The Himalaya is synonymous of Lord Shiva for Hindus. Circumambulation of Kailash-Mansarover, Char Dhams, and Shakti Peets in the Himalaya is century-old tradition, which promotes pilgrimage tourism in the Himalaya. The natural beauty of the Himalaya is immense. The snow-capped mountain peaks, alpine pasturelands, river valleys, terraced agricultural fields, and numerous topographies make the Himalaya as a natural destination. The Himalaya supports to the development of adventure tourism in the form of mountaineering, river rafting, and trekking as it has numerous destinations. The forest landscape is unique.

The mountain peaks, river valleys, and wilderness make the Himalaya panoramic, whereas these entities can be reached by heart rendering trekking with full of enthusiasm. Reaching to the Himalaya needs a long and day and night unbreakable journey, it may take years. Before trekking to the Himalaya, one has to understand it, its height, solace, divinity, and resources. Dhaula Dhar Range of Himachal Pradesh with gentle slope and Markha Valley of Ladakh with challenging routes are the major trekking destinations. Mountaineering in many mountain peaks like Nanda Devi, Kangchenjunga, and Shigri Parbat are the most popular. Zanskar, Beas, Teesta, Ganga, Tons, Kali rivers are the centers of white-water rafting and kayaking. Biking is practiced in Ladakh, Uttarakhand, Himachal Pradesh, Sikkim, and Arunachal Pradesh. Paragliding has been developing as an adventure tourism in the Himalaya. Skiing is practiced in Gulmarg in Kashmir and Auli in Uttarakhand. There are thousands of national parks, wildlife sanctuaries, biosphere reserve, and conservation parks, where eco-tourism is practiced. Tourism is one among the sources of livelihood of the people in Himalayan region and it shares more than 50% of the revenue of Ladakh region, Uttarakhand, Nepal, Bhutan, and Tibet.

Himalaya has a few mineral deposits, some of which are sapphires in the Zanskar Range, alluvial gold in the Indus River, copper ore in Baltistan, and iron ore in the Vale of Kashmir. Ladakh possesses borax and sulfur, coal in Jammu Hills, and Bauxite in Kashmir. Nepal, Bhutan, and Sikkim have coal, mica, gypsum, and graphite, and ore of iron, copper, lead, and zinc. Meanwhile, the Himalaya may have a vast reservoir of petroleum.

1.9 A Thousand Trails

Himalaya has thousand trails. In the past these trails and footpaths were the main means of communications in the Himalaya. Now, the roads are constructed in many areas and it has become the major means of transportation. In addition of roads, some areas are connected by rail routes and by airways. There are many passes which connects the remote mountain regions and the Himalayan countries. Kodari Pass connects Nepal with Tibet. Shipki Pass is located in the India-Tibet border area. Manali-Leh National Highway crosses the Great Himalayan Range. Leh is also connected by the Zanskar Range and the Vale of Kashmir. Khardung Pass (5404 m) is the highest place, where the national highway is constructed. Lipulekh Pass connects Pithoragarh (Uttarakhand) with Tibet and Nepal. Mana and Niti Passes connects Chamoli district with Tibet. Zoji La Pass connects Jammu and Kashmir with Ladakh. Nathu La Pass connects Tibet with Sikkim. There are a thousand of trekking routes, which connect the various regions with the Himalayan countries. Trekking to the Himalayan pilgrimages are only the routes to reach in the past and still now, many Himalayan pilgrimages are reached only by trekking.

The Sindhu and Sutlej are the transboundary rivers among India, Tibet, and Pakistan; the Brahmaputra Rivers makes transboundary among India, Tibet, and Bangladesh; the Ganga makes transboundary between India and Bangladesh. There are many tributaries of the Ganga, Brahmaputra, and the Sindhu Rivers which make transboundary among the South Asian countries. The Lesser Himalaya and Tarai regions are connected by railways. Kalka-Shimla and Siliguri-Darjeeling are the two mountainous narrow gage railway lines. Jammu, Pathankot, Dehradun, Haridwar, Kotdwar, and Kathgodam cities are also connected by railways. Srinagar, Leh, Dehradun, Rudrapur, Kathmandu, Pokhara, Paro, and Lhasa are well connected by airways. In the past, traders, shepherds, and pilgrims undertook the journey through the Himalaya. Journey through the Himalaya is believed to be the means of liberation from the cycle of birth and death.

1.10 Culture, Customs, and Rituals

Beyond its natural treasures, the Himalaya holds a rich cultural heritage and spiritual significance (Kandari and Gusain 2001). On the other hand, the Himalaya is believed to be an abode of thousands of folk deities, it is believed to be the embodiment of Lord Shiva and Shakti, the Hindus adorable folk deities, on the other. All the natural entities—land, water, and forests are considered the folk deities and these are worshipped on various occasions. The Ganga River is called the mother Ganga and taking a holy dip in the Ganga is believed to liberate a soul from the cycle of birth and death. The Yamuna River is renowned as the sister of Yama and taking a ship of water from it makes humans disease frees. The rivers Saraswati, Alaknanda, Bhagirathi, Vishnu Ganga, Mandakini, Nandakini, and several others are called after the folk deities (Swami 2001. The Brahmaputra River is believed to be the son of the Brahma, one amongst the three Hindu deities. The mountain peaks are also named after the folk deities. The Nanda Devi Mount after the folk deity Nanda, The Trishuli Mount symbolizes the Trishul of Lord Shiva, the Annapurna Mount is named after the Goddess Annapurna, who fulfills food to people in the world. Mount Dronagiri also has cultural and historical importance. These rivers and mountain peaks are believed to be the Gods and Goddesses (Ramanan 2017).

The Himalaya is renowned for its highland sacred locales, which are called the highland pilgrimages. These highland pilgrimages are sacred mainly for the followers of Hindus, Buddhists, Jains, and Sikhs. The famous four Dhams of

Uttarakhand—Badrinath, Kedarnath, Gangotri, and Yamunotri (located above 3000 m), and Kailash Mansarover (located above 4000 m) are the main highland pilgrimages for the Hindus, Jains, and Sikhs. For the followers of Sikhism, Govind Ghat and Hemkund Sahib are the popular highland sacred places. In Ladakh and Bhutan, there are many sacred monasteries, where the Buddhists perform Puja of Lord Buddha. The two sacred monasteries of Hemis and Thiksey are located in Ladakh. The river valley pilgrimages are many among which, the two pilgrimages—Rishikesh and Haridwar are famous. Rishikesh is known as the "World Capital of Yoga" and Haridwar/Hardwar is known as the gateway to Badrinath and Kedarnath, respectively. In the Indian Central Himalayan Region, Devalaya, Shivalaya, Shakti Peeths, Siddha Peeths, and Jyotirlinga are the major sacred places where pilgrims visit and perform Puja throughout the year.

The Himalaya provides a tranquil and spiritually charged environment for meditation, self-reflection, and inner exploration. They uphold traditional culture and society (Manhas et al. 2019). Colorful festivals, vibrant folk music, and traditional crafts make the Himalaya unique and rich in culture and customs. The people of the Himalayan region celebrate fairs and festivals throughout the year. The first day of every month, according to the Hindu calendar, is known as Sankranti, and every member of society celebrates it. The fairs and festivals are organized according to the seasons and climate and are closely related to nature. Each year, the Navratris of Chaitra and Ashwin are celebrated continuously for nine days and nights. During these festivals, people worship various forms of Durga, the folk deity. In many areas, the *Dev-Devi Naach* is performed, during which the folk deities are believed to appear in human bodies, known as "Dev Avataran," and the humans dance. This practice is called *Dev-Devi Naach.* Rai Sankranti (Harela) is associated with the plantation of trees and the conservation of nature and is celebrated in Shravan. Similarly, Ghee Sankranti, celebrated in Bhadra, is linked to animal herders and their prosperity. During this festival, the animal herders play a Holi-like celebration with buttermilk, curd, and ghee in the alpine pasturelands, mainly in Uttarakhand. Makar Sankranti, Uttarayani, and Basant Panchami are celebrated with great enthusiasm, and fairs are organized in various places and along the river valleys. One of the most famous festivals is Vaisakhi, celebrated on the first day of Vaisakh and continuing for 5 days. During Vaisakhi, idols of folk deities are taken in a cultural procession, given a holy dip in the river, and then returned to their place, an event known as *Dev-Sthan.* Festivals are also celebrated based on the Earth's revolution and the Sun's position in the northern and southern hemispheres.

The *Solah Samskaras* are performed from the birth of a child until their death. These ceremonies begin with the *Namkaran* (naming) ceremony and continue one by one (Encyclopaedia Britannica 2025). Other main *Samskaras* include *Annaprashan* (first feeding of solid food), *Aksharambh* (initiation into learning), *Yagyopaveet* (sacred thread ceremony), *Vivah* (marriage), and *Antyeshti* (funeral rites). People worship various folk deities, as well as rivers, forests, mountains, jungle deities, spirits, and *Veers* (brave individuals). They also perform rituals to honor the land before sowing seeds, after harvesting crops, and during the foundation ceremonies for houses, cowsheds, and other economic ventures. Water sources such as natural springs, streams, rivulets, and rivers are worshipped on various occasions. During the summer season, people worship the God of Rain, *Latu Devta*, to seek sufficient rainfall, particularly for sowing seeds and mitigating drought situations.

The book "Exploring the Mystical Himalaya: A Thousand Trails" is based on my visits of the different parts of the Himalaya—Ladakh, the Indian Central Himalayan Region, Nepal, Bhutan, and Tibet. It took more than two decades to travel the Himalayan regions. In these regions, I visited many urban and rural settlements and natural and cultural locales several times. I spent times with the villagers and interviewed them on different culture, customs, and rituals, the people of the Himalayan region performed. The purpose of composing this book was to explore the mystical Himalaya, its nature and culture. It includes nine chapters, including the introduction and conclusions. The others are Hari Ki Dun: The Valley of Lord Vishnu, The Legends of Kedarkhand: A Journey through the Himalayan Abode, Manaskhand: The Climax of Nature and Culture, In the Land of Lamas, Kathmandu to Mukhimath: Unity and Continuity through the Kali Gandaki River, Harmony in the Himalaya: Buddhism and Bhutan, and Beyond Borders and Culture. The book is immensely important to all the stakeholders—students, scholars, academicians, tourists, and pilgrims.

1.11 Organization of the Book

This book "Exploring the Mystical Himalaya: Cultures, Economy, and Environment" contains a total of nine chapters including the introduction and conclusions. Each chapter has its significance and uniqueness as it illustrates about the difference aspects and the regions of the Himalaya. Here, I have elaborated about each chapter separately. The introduction is the first chapter, which includes landscape and landforms; Himalayan climate and its implications; Himalayan glaciers and rivers; fauna, flora, and avifauna; national parks and wildlife sanctuaries; hydropower projects; social systems; economy and development; a thousand trails; and culture, customs, and rituals.

Chapter 2 "Ladakh: The Land of Lamas" is illustrated. In August 2022, I visited Ladakh region, starting the journey from Leh to Nubra, and Pangong including Kargil. The Ladakh is a Trans-Himalaya, a cold desert, and a tableland (plateau) with altitude ranges from 3000 m to more than 6000 m. During the summer, the river valleys—Shyok, Nubra, and Sindhu remain hot. Large sand dunes are found in Nubra Valley with double humped camels. On the other hand, Pangong is the highest lake in the world, which is the largest and occupied by both India and China. Ladakh is the land of Lamas and the people follow Buddhism. Ladakh has many world-famous monasteries. The second largest population belongs to Muslim. Ladakh is surrounded by Pakistan and China therefore it has high geostrategic importance. In this chapter, the culture, customs, and ritual of Ladakh have been described. In addition, the occupation of Ladakhi people, mainly agriculture and cow herding, has been illustrated largely.

Chapter 3 illustrates the Hari Ki Dun: The Valley of Lord Vishnu. The chapter is based on my visit to Hari Ki Dun in July 2024. The journey covers approximately 230 km distance from Dehradun to Hari Ki Dun via Mussoorie, Yamuna Bridge, Nainbag, Naugaon, Purola, Mori, Netwar, Sankari, Dharkot, Gangad, and Osla along the Yamuna River, the Kamal River, and the Tons River, respectively. In this chapter, natural and cultural aspects of the Hari Ki Dun valley are described. Altitude varies from about 600 m to more than 3000 m. The entire route is characterized by three sub-cultural regions—Garhwal, Jaunsar-Bawar, and Rawain. The natural landscape, forest landscape, and agricultural landscape change from the river valleys to middle altitudes and the high Himalaya. The terraced agriculture is practiced with potatoes grow in the lower elevation, a variety of fruits, including apple, and vegetables grow in the middle altitudes, and potato and Rajma grow in the highland areas. The change in cropping patterns is due to climate and climatic zones—subtropical, temperate, and cold. It also elaborates culture, customs, rituals, fairs, and festivals of the Tons River valley and Hari Ki Dun.

Chapter 4, "The Legend of Kedarkhand: A Journey Through the Himalayan Abode," is described. This includes a journey from Haridwar to Badrinath and Kedarnath, journey from Rishikesh to Gangotri and Yamunotri, a journey from Karnprayag to Homekund, Sutol, and Nand Prayag, and a journey from Haridwar to Kotdwar, Pauri, and Srinagar. In these routes, a detailed description of culture, society, customs, rituals, and nature was presented. Further, the occupation and livelihood options of people in these routes were described. It was noticed that the culture, customs, and rituals vary from one route to another. Kedarkhand is the western part of the Uttarakhand Himalaya and its history is ancient. In this part of Himalaya, many highland pilgrimages including Panch Badris, Panch Kedars, and Panch Prayags are located. In addition, two world-famous valley pilgrimages are located in this cultural realm of the Uttarakhand Himalaya. It is the home for the major river systems of India. All these descriptions are presented in this chapter. I traveled these routes several times.

"Manaskhand: The Climax of Nature and Culture" in the Chap. 5, which is divided into two parts. Manaskhand is also known as the Kumaon Himalaya. The first one is the description of the Western Kumaon and the second one is the description of the Eastern (Kali) Kumaon. The first part starts from the border areas of the Garhwal Himalaya. There are mainly three entrances to the Kumaon Region from the Garhwal region. These are Gwaldam, near Karnprayag, Pandu Khal near Gairsain, and southeast part of Pauri district. While my trekking to this part I covered Bageshwar, Almora, and Nainital districts and their famous natural and cultural destinations. The cultural destinations in this part include Bageshwar, Jageshwar, Sun Temples of Katarmal, Kasar Devi Temple, Naina Devi Temple, and Golu Devta Temple. In the natural locales, Kosani, Almora, Nainital, and seven taal (lakes) located in Nainital district. In the eastern part, the famous natural destinations are Munsiyari (the Switzerland of Uttarakhand), Champawat, Pithoragarh, and Dharchula. In terms of cultural locales, Om Parvat, Parvati Kund, and Adi Kailash are famous. Three valleys are famous in this route, which are the Johar Valley (Gori Ganga), the Darma Valley (Dhauli Ganga E), and the Vyas Valley (Kali River). I have traveled these two parts of the Manaskhand during the two different times. The western part of Manaskhand, I visited in 2022 and in the eastern part, I visited in 2024. I trekked to the Johar Valley, Darma Valley, and the Vyas Valley during the winter, accordingly wrote travelogs, and incorporated it in this book.

I had an illustrated and long visit from Kathmandu to Pokhara and thereafter a trekking to Muktinath (Upper Mustang Valley) in October 2023. Chapter 6 is devoted to "Kathmandu to Mukhimath: Unity and Continuity Through the Kali Gandaki River." In this chapter, a detailed description of culture and nature along the Kathmandu and Pokhara by road and Pokhara to Mukhimath by trekking is presented. Kathmandu is a capital city of Nepal and it is culturally rich city. The famous Temple of Pashupati Nath, one of the Jyotirlingas, is located on the bank of the Baghmati River in Kathmandu. The Baghmati River is considered as the holy river of Nepal, which Hindus perform all their rituals. On the other hand, Pokhara is a city of natural beauty and known for its underground rivers. Besides, the city is also considered as a cultural city. A trek route to the famous highland pilgrimage "Muktinath" starts from Pokhara along the Kali Gandaki River. The Kali Gandaki River is another holy river of Nepal, which is worshiped as the Ganga is worshiped in India. The descriptions of cultural and natural characteristics of these two routes are depicted in this chapter.

Harmony in the Himalaya: Buddhism in Bhutan, in the Chap. 7, based on my visit to Thimphu and its surrounding areas from 31 March to 4 April 2024. I landed at the International Airport, Paro, Bhutan and then traveled to Thimphu by road. Traveled in many cultural and natural places, I interviewed the priests and old and experienced people. Bhutan is only a country in the world where almost all people follow Buddhism. They live in harmony with the nature. The Bhutanese are very peaceful in nature and they are quite satisfied with their economy and income, and therefore, the happiness index of Bhutanese is the highest in the world. Himalaya covers more than 30% of the total area of Bhutan and it is known as the Bhutan Himalaya. Bhutan has spectacular natural, forest, and cultural landscapes. The chapter is divided into two parts. The first part illustrates geography, economy, culture, and population and the second part is a detailed description of my travelog.

I had many visits to Tibet and China at least for a decade. During the period, while staying in Chengdu, a capital city of Sichuan, I have visited many areas of Tibet and its surrounding areas, among which, Lhasa, Moxi, and Mount Gonga were prominent. Tibet is a highest tableland (plateau) in the world and it is a cold desert. People of Tibet follow Buddhism and their cultural heritage is rich. They are peaceful people and their cultural relation with India is unique and integrated. Chap. 8, "Beyond Boarders and Culture" deals with geography, economy, and culture of Tibet. A historical description is presented and how Tibet became the autonomous region of China is described in this chapter. The last Chap. 9 is the conclusions, which is summary of the all chapters.

References

Aitken B (2003) Footloose in the Himalaya. Permanent Black, Delhi

Aryal S, Ghimire S, Tiwari S, Baaniya Y, Pandey VP (2024) Evolution and future prospects of the hydropower sector in Nepal: a review. Heliyon 10(10):e31139. https://doi.org/10.1016/j.heliyon.2024.e31139

Belho K, Rawat MS, Rawat PK (2024) Anthropogenic climate change accelerating monsoon hydrological hazards in Northeastern Himalayan region of India: geospatial approach. Arab J Geosci 17(2). https://doi.org/10.1007/s12517-024-11871-1

Bijalwan A, Mamgain A, Rathaude SK (2018) Crop-biodiversity along altitude in Gangotri valley of Uttarkashi district of Garhwal Himalaya, India. Biodivers Int J 2(4):378–383. https://doi.org/10.15406/bij.2018.02.00089

Chand S, Mishra L (2024) Pilgrimage tourism in Uttarakhand: a critical review with reference to the char Dham Yatra. Int J Res Publ Sem 15(1):281–287. https://doi.org/10.36676/jrps.v15.i1.1647

Choudhary DK, Garkoti SC (2023) Transhumant pastoralism in Indian western Himalaya: traditional ecological knowledge and contemporary practices. Int J Sustain Dev World Ecol 31(1):57–70. https://doi.org/10.1080/13504509.2023.2254273

Desai A, Mukhopadhyay I, Ray A (2023) Solar PV–hydropower enhanced picogrid as a sustainable energy model for hilly remote areas. Energy Inform 6:Article 24. https://doi.org/10.1186/s42162-023-00294-z

Devi NB, Lepcha NT, Mahalik SS, Dutta D, Tsanglao BL (2021) Urban sacred grove forests are potential carbon stores: a case study from Sikkim Himalaya. Environ Challenges 4:100072. https://doi.org/10.1016/j.envc.2021.100072

Encyclopaedia Britannica (2025) Samskara. In Encyclopaedia Britannica. https://www.britannica.com/topic/samskara-Hindu-passage-rite. Accessed July 2025

Guha R (2000) The unquiet woods: ecological change and peasant resistance in the Himalaya (expanded edition). University of California Press, Berkeley

ICIMOD (2024) Mountain agriculture systems in the Hindu Kush Himalaya [backgrounder]. International Centre for Integrated Mountain Development (ICIMOD). https://www.icimod.org/mountain/agriculture/

Joshi VC, Sundriyal RC, Arya D (2025) Impacts of climate warming on biodiversity, ecosystem functionality, and services in the Himalayan landscape. In: Climate change impact on Himalayan biodiversity. Springer, Cham, pp 337–351. https://doi.org/10.1007/978-3-031-77149-1_14

Kandari OP, Gusain OP (eds) (2001) Garhwal Himalaya: nature, culture & society. Transmedia, Srinagar

Kulkarni AV, Karyakarte Y (2014) Observed changes in Himalayan glaciers. Curr Sci 106(2):237–244

Manhas RK, Charak S, Sharma A (2019) Exploring the dimensions of spirituality, wellness and value creation amidst Himalayan regions: promoting entrepreneurship and sustainability. Indian J Tour Sustain 7(2):45–63. Retrieved from ResearchGate, 2023

Mishra H, Pandey BW, Mukwada G, de los Ríos P, Nigam N, Sahu N (2023) Trapped within nature: climatic variability and its impact on traditional livelihood of Gaddi transhumance of Indian Himalayas. Local Environ 28(5):547–563. https://doi.org/10.1080/13549839.2022.2162025

Nandy SN, Dhyani PP, Samal PK (2006) Resource information database of the Indian Himalaya. GBPIHED, Almora

Negi VS, Pandey A, Singh A, Bahukhandi A, Pharswan DS, Gaira KS, Wani ZA, Bhat JA, Siddiqui S, Yassin HM (2024) Elevation gradients alter vegetation attributes in mountain ecosystems of eastern Himalaya. India Front For Glob Change 7:1381488. https://doi.org/10.3389/ffgc.2024.1381488

Oli KP, Pandey MR (2024) The horizon of the third pole: mapping future scenarios and strategic responses. Himalayan J Environ Zool 15(2):112–135. https://doi.org/10.1177/18785395241293282

Parashar D, Sarkar MS, Singh AP, Abdulqadim TJ, Tolche AD, Khan MYA (2024) Spatiotemporal snowline status and climate variability impact assessment: a case study of Pindari River basin, Kumaun Himalaya, India. Environ Sci 36(104). https://doi.org/10.1186/s12302-024-00924-7
Ramanan V (2017) The legend behind Nanda Devi mountain. The Hindu https://www.thehindu.com/society/history-and-culture/the-legend-behind-nanda-devi/article17475603.ece
Rao MBR, Shukla UK, Raju NJ (2008) Geomorphology of Ganga Basin and channel planform dynamics. Int J Sed Res
Sale R, Cleare J (2000) Climbing the world's 14 highest mountain. The Mountaineers, Seattle
Sati VP (2014) Landscape vulnerability and rehabilitation issues: a study of hydropower projects in the Garhwal region, Himalaya. Nat Haz 75(3):2265–2278. https://doi.org/10.1007/s11069-014-1430-y
Sati VP (2019) Himalaya on the threshold of change. Springer, Cham
Sati VP (2021) The Ganges: cultural, economic, and environmental significance. Springer, Cham, p 240
Sati VP (2022) Uttarakhand: society, culture, and pilgrimages. Today and Tomorrow, Printers and Publishers, Delhi
Sati VP (2023) Natural and cultural diversity in the Himalaya. Springer, Cham
Sati VP (2024) Exploring Uttarakhand's temples and divine paths: sacred sanctuaries. Cambridge Scholars Publishing, London
Sherpa C (2025) Non-timber forest products and their role in rural livelihoods: a case study from the Annapurna conservation area, Nepal. Ethnobot Res Appl 30:1–15
Shrestha AB, Agrawal NK, Alfthan B, Bajracharya SR, Maréchal J, van Oort B (eds) (2015) The Himalayan climate and water atlas: impact of climate change on water resources in five of Asia's major River basins. ICIMOD, GRID-Arendal and CICERO, Patan, p 58
Singh S, Kerven C (2023) Pastoralism in South Asia: contemporary stresses and adaptations of Himalayan pastoralists. Pastoralism 13(1):21. https://doi.org/10.1186/s13570-023-00283-7
Swami TM (1960) Wanderings in the Himalayas, English edition, Madras, Chinmaya publication trust, 1960. Translated by T.N. Kesava Pillai
Swami S (2001) Himalaya: through the lens of a sadhu. Tapovan Kuti Prakashan, Uttarakashi
Valdiya KS (2016) The making of India: geodynamic evolution. Springer, Cham. https://doi.org/10.1007/978-3-319-26058-0
Wester P, Mishra A, Mukherji A, Shrestha AB (eds) (2019) The Hindu Kush Himalaya assessment: mountains, climate change, sustainability and people. Springer, Cham
WWF (2024) Kangchenjunga conservation area survey reveals key data on snow leopard populations in Nepal. https://www.wwfbhutan.org.bt/?388476/Kangchenjunga-Conservation-Area-Survey-Reveals-Key-Data-on-Snow-Leopard-Populations-in-Nepal

Chapter 2
Ladakh: The Land of Lamas

Contents

Abstract Ladakh, a union territory of the Republic of India, is located in the Trans-Himalayan region and is known as the "Land of the Lamas." It is a cold desert characterized by its stony and barren landscape. Inhabited by people of all three major religions—Buddhism, Islam, and Hinduism—it boasts a rich and diverse culture. The climate remains very hot during the summer and extremely cold during the winter. Ladakh is a tableland, or a highland plateau. Its economy depends primarily on tourism and livestock rearing. A few fruits, vegetables, and grain crops are grown along the river valleys—mainly the Sindhu, Zanskar, and Shyok—but production remains very low. This chapter provides a detailed description of the land, people, culture, and economy, supported by the latest illustrations.

Keywords Lamas · Sindhu river · Shyok river · Zanskar river · Pangong lake · Lamas

V. P. Sati, *Exploring the Mystical Himalaya*, Springer Geography,
https://doi.org/10.1007/978-3-032-14275-7_2

2.1 Introduction

Ladakh, located in the Trans-Himalaya, is a union territory of the republic of India. It has peculiar climate, in terms of a long and frigid winter, however, the 2 months—July and August are the most suitable time due to two reasons. The first one is that, during the other months of the year, Ladakh remains snow-capped. The second one is that Ladakh is a rain-shadow region. When the rest of India receives heavy downpours during the monsoon, the Ladakh region remains dry. Therefore, Ladakh is the highest cold desert in the world, where altitude begins at 3000 m in Leh and rises above 7000 m, with its latitudinal extension above 38° (Hayward 1870). During this time, solar radiation falls vertically and intensively, which often leads to sunburns and affects the eyes (Fig. 2.1). A clear sky, mainly during the summer, accelerates the intensity of solar radiation. On the other hand, during winter, the temperature in the Ladakh region drops to −20 °C, and it receives intense snow. However, the snow-melting rate is higher in areas below 3500 m, and after a maximum of 2 days of snowfall, it starts melting (Kale 2014). Leh city, situated at an average altitude of 3400 m, receives snow during the winter but does not retain it, even when the temperature is below freezing.

As Ladakh is a dry reason, it has very sparse natural vegetations. The mountain peaks are snow-capped where the slopes steep and gentle are stony, baren, and deserted. The forests are found in the river basins, mainly along the Sindh River and made oasis. The vegetation is typical, forms conical shape. The principal species are

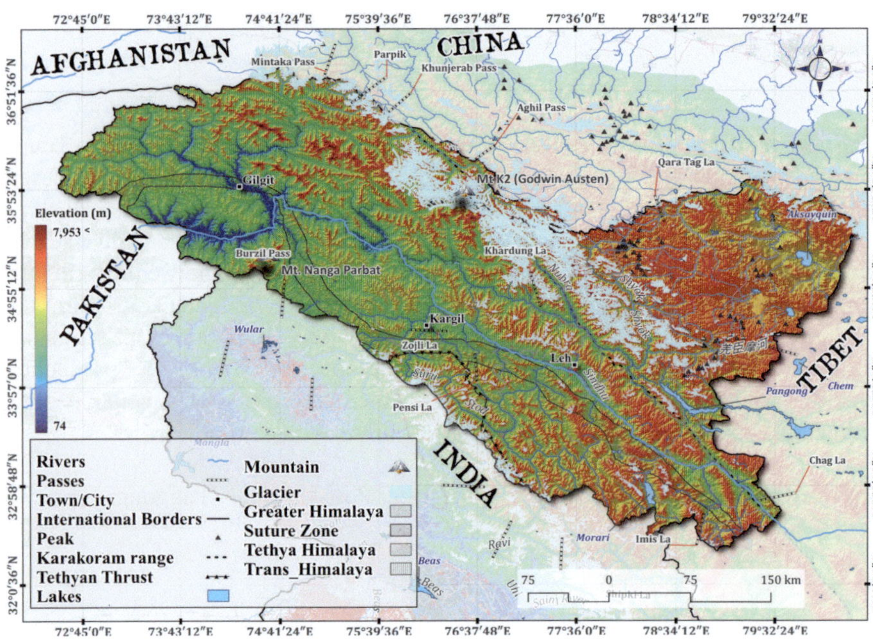

Fig. 2.1 Location map of Leh Ladakh union territory in India

poplars, willows, and patches of grasses and shrubs. Among the fruits, apricots and walnuts are prominent. This region holds a rich faunal diversity among which, yaks, ibex, Tibetan hare, blue sheep, wild ass, Bharal, Marmot, the Tibetan antelope, and the Snow leopard are the principal ones. It has 310 bird species, and some of them are Golden Eagle, Wall creeper, Rock Bunting, black-necked crane, and Himalayan Snowcock. Ladakh has many national parks and wildlife sanctuaries—Hemis National Park and Changtang Wildlife Sanctuary are renowned (Dar et al., 2025; Lamo et al., 2024). Furthermore, about 1100 medicinal plants are recorded in Ladakh (Jina 1996). Therefore, it is called a paradise for wildlife enthusiasts. These sanctuaries provide suitable bases for the promotion of eco-tourism in the region. Already, the Nubra Valley is promoting eco-tourism in Ladakh.

The Ladakh has many glacier-fed perennial rivers with fresh water. During the winter, these rivers remain frozen mainly in the higher reaches whereas, in the summer, these rivers inundate and causing flash in the downstream areas. In several occasions, there are flash floods in the rivers. As the roads leading to the many popular destinations, constructed along the course of these rivers, these roads are flooded during the summer, and many occasions, it leads to damage property and human casualties. This happens mainly between the Nubra and Pangong Lake along the Shyok River. The major rivers are Sindhu, Zanskar, Shyok, Nubra, Sumdah, Markha, and Rumbak. Some beautiful lakes are Pangong, Tso, and Moriri Tso. These rivers and lakes make spectacular river valley and lakes landscapes.

Ladakh is situated in the Trans-Himalaya, bordered by Tibet (China) in the east, Pak-Occupied Kashmir in the North, Jammu, and Kashmir in the west, and Himachal Pradesh in the south. It covers approximately 117,000 km^2 area. Ladakh Range which is a southeastern extension of the Karakoram Range and upper Indus River valley are the major landforms of Ladakh. It is a vast area, located in the northern and eastern part of the Kashmir region. A large part of Ladakh is illegally occupied by Pakistan in the north and China in the east. The northern part is known as Gilgit-Baltistan and the eastern part is known as Aksai Chin. Ladakh is the highest table land, a cold desert, characterized by mountains, plains, and deep valleys. Altitude starts from around 3000 m and more than 5000 m. Rupshu is an area located in the southeastern Ladakh, which have brackish lakes with a uniform altitude of about 4100 m. Zanskar Range lies in Ladakh forms inaccessible region because of steep slopes and snow. In this region, people and cattle remain indoor because of cold climate. The Zanskar River flows here, which joins the Sindh River in proximity of Leh.

The climate of Ladakh is dry and cold. However, it is very hot during the summer season, when temperature exceeds 27 °C. The Nubra Valley is the hottest, where sand dunes are found along the Shyok River. It is characterized by snow in the peaks of high mountains and deserts in the valleys. Annual average rainfall is very nominal, approximately 80 mm. Vegetation is confined to the river valleys where water is available. These areas make green oasis. The plant species are tamarisk, shrubs, and furze. The Sindhu River flows about 50 km in Ladakh, makes a beautiful landscape between Spitok Gompa (near Leh airport) and Hemis Gompa in the heartland of Ladakh.

2.2 History and Culture

Ladakh is a land of high passes and renowned for trekking and bike riding. It was known as Maryul means lowland and others called it Kha-Chumpa. It was referred as Kia-Chha and Ma-Lo-Pho in the writings of Fa-Hein and Hiuen Tsang, respectively. The first immigrants in Ladakh and the Indus Valley are believed to be the Brokpas of Dadarstan and they are popularly known as Sham. The others came from Karja (Kulu) were the Mons and Aryans, who first settled in Gya and then in Rong, Shyok, Sakti Tangtse, and Durbuk. Dard, an Indo-Aryan race were the ancient inhabitants of Ladakh (Mann 2002). Later on, Buddhism spread from India to Tibet via Ladakh, which leaves its imprints on this land. Islamic, German, and British missionaries made inroads toward conversion but little success.

Ladakh was invaded by a Tibet ruler Nemagon in tenth century AD. During the period, Ladakh was divided into small principalities and fighting with each other. Memagon defeated them one by one and he established his kingdom named Shey, 15 km from Leh (Kashmir 2006). It was an independent country since then. The successor of Nemagon named King Singge Namgyal consolidated the Ladakh Empire and made it a strong kingdom. He was a monarch, a statesman, a diplomat, and a builder. He built a beautiful and historic nine storied Leh palace. He was the pioneer of promoting horse polo in Ladakh.

The present Leh district was a part of Greater Ladakh in the ancient times, which extent from Kailash Mansarover to Swaat (Dardistan). It was independent state without any influence of Tibet. With the war between China, Tibet, and Baltistan, the Greater Ladakh divided into several parts in the eighth century AD. After India's independence, Pakistan and China captured 78,114 km^2 and 37,555 km^2 land of Ladakh in both sides, respectively. About 5180 km^2 land, Pakistan gifted to China, illegally, comprising the parts of present Leh and Kargil districts. Ladakh was divided into two districts—Leh and Kargil in 1979. On 11 December 2023, after abolition of 370 from Jammu and Kashmir State, Ladakh was declared as the Union Territory of the Democratic Republic of India.

Ladakhi people have a vibrant and immersive culture. They are very interesting and welcoming. The visitors participate in the folk dance and song during various occasions. Ladakhi culture is influenced by Tibetan culture as both areas are Trans-Himalaya and having similar landscape and climate. The Buddhist tradition from India also influences Ladakhi culture. Woolen clothes are mainly woven by local people using traditional weaving methods. They wear quite different dresses. Men wear long woolen robes called Goucha. Women also wear similar robe with different style, which is called Kuntop and the Bok. Kabra, Khala, Porokshkang, Sagam, Shangho, and Solo are popular Ladakhi dishes. Thukpa is the most popular dish, which comprises soupy noodle dish assorted vegetables and minced meat. Other dishes include Tigmo, Momos, and Noodles.

Ladakhi perform different festivals such as Hemis, Tsechu, and Losar. They worship folk deities, sing sons, perform mask dance, exchange gifts, and play camel racing and archery. The nature of festivals changes according to changing seasons.

The Saka Dawa Festival is unique, considered to be the holiest Buddhist vacation. A festival, which is celebrated every fifth month of the Tibetan lunar calendar is known as Phyang Tsedup Festival. A very important festival is Sindhu Darshan festival, celebrated on the auspicious day of Guru Purnima on the bank of the Sindu River close to Leh. A festival is celebrated in cave Gompa of Tak-Tok is called Tak-Tok festival. Dosmoche is another popular festival celebrated in a year.

Ladakhi people are nature lovers. They perform dances that align with the mood of nature. The landscape varies, and accordingly, the dances are shaped. The most popular dance is called Khatok Chenmo. This traditional dance is performed during family functions and festivals. The people who perform it are known as Shandal. Among the other dances are Kompa, Tsum-tsak, Jabro, Chaams, Chabs-Skyan Tses, Raldi Tses, and Alley Yaato. One unique feature of Ladakh's dances is that the music is either in Sanskrit or Tibetan, and they chant in both languages. The musical instruments used are traditional, such as the Linyu (flute), Damnyan (string instrument), Piwang, Khakong (sitar), Daph (dafli), Daman, Surna, and Piwang (shehnai and drum).

They practice the polyandry system, which is borrowed from Tibet. In this system, a woman has two or more husbands. Weddings are long celebrations. Village families bring food to feed the people involved in the ceremony, including the groom and bride. The groom often drinks Chang and dances, whereas the bride sits silently in the kitchen. The horoscopes of both the groom and bride are matched to determine whether the marriage is possible. There is also an interchange in marriage between two families. For instance, a brother and sister from one family can marry the sister and brother from another family, respectively. This practice is common in the Changthang community. If family members do not agree to the marriage of a young couple, the couple may elope. Recently, under a new government scheme, the government provides one-time financial assistance of Rs. 50,000 to female beneficiaries who are economically weak.

Monasteries, also known as gompas, are located in every village and settlement. These are centers of worship, often built on hillsides with massive walls, stupas, and prayer flags (Toran Patake). The walls are painted with illustrations of the Buddha and his disciples. In Ladakh, some monasteries date back to the eleventh century. The architecture of Ladakh combines Indian and Tibetan monastic styles. There are ten famous monasteries in Ladakh that represent the living heritage of the Buddha. These include Hemis, Thiksey, Diskit, Likir, Spituk, Stakna, Alchi, Phyang, Takthok, and Sumda Chun monasteries.

In Buddhism, worshipping Buddha is performed in groups for the attainment of spirituality and moral well-being. It is also done to appease evil spirits. At the place, where chanting is performed, folklore—dance and music is also performed. The monks wear special costumes and make hand gestures. Ladakh is an important Buddhist Center. Buddhist came here in the first century and in the second century, Buddhism spread in Tibet. Buddhism flourished in Ladakh and the people continued following the Buddhism pathway although later on, Islam also entered in the land and now the Buddhist population is 70%. Tourism in the course of time became the major source of economy and the option of livelihood and now the economy of Ladakh is mainly rely on tourism.

2.3 Agricultural and Livestock Practices

Agriculture is the backbone of Ladakh's economy as about 70% working force are marginal farmers. It is practiced along the Sindh River, wherever, the land is plain and fertile. These areas are accessible. The upland valleys, where shepherds migrate to rear goat, sheep, lambs, and horses, are not fit for faming agricultural crops. Agriculture, horticulture, and pasturing animals are the farming systems. Mainly monocropping is practiced. However, double cropping is practiced in Sham belt and Nubra Valley. The agriculture season is very short in Ladakh even during the summer, it is difficult to grow crops. Some areas, vegetables and fruits are grown whereas the quantity is less. Among the agricultural crops wheat (autumn), barley, millets, buckwheat, peas, beans, and turnips are prominent crops. Farmers also grow alfa-alfa and fodder. Among fruits, apricot, apples, pear, and peach are grown. Colliflower, green leave, carrot, radish, and potato are grown (Tewari et al. 2015). The farmers practice organic farming. Agriculture is traditionally practiced whereas with the advent of agricultural department and some technology institutions established in Leh and the farmers using technology. Some farmers are using tractor for ploughing crop fields and making polyhouses for growing vegetables.

Ladakh has a very hostile climate for crop cultivation and the practice of sustainable agriculture. The region remains cold and dry during winter and hot and dry in summer. Agriculture is practiced only for a few months in summer, with very low output. Recently, the Agricultural Department of Ladakh has introduced new innovations, such as providing greenhouses to farmers to grow various crops in accordance with the prevailing climate. This initiative has shown positive results in sustaining the livelihoods of farmers in Ladakh (Down to Earth 2025). Scientific innovations, particularly low-cost sustainable technologies for winter vegetable cultivation in the lower belt of Ladakh, are gaining momentum and have proven successful (Dolkar and Mehdi 2025). The cultivation of medicinal and aromatic plants is also increasing in the region. The Council of Scientific and Industrial Research (CSIR) is providing extensive support to scientists working in Ladakh and has already demonstrated the feasibility of such cultivation (Ladakh Autonomous Hill Development Council (LAHDC) 2024). Grape cultivation is a new initiative in Ladakh, supported by the Indian Council of Agricultural Research (ICAR). Recently, grape farming has expanded in the region (Indian Council of Agricultural Research (ICAR) 2024). The Agricultural Department, backed by the Union Territory Administration of Ladakh, is now focusing on agricultural development in the context of climate change. The Central Arid Zone Research Institute (CAZRI) is also contributing by organizing workshops and seminars and engaging with the local community and farmers (Reach Ladakh Bulletin 2025). Tomato is another crop with high potential in Ladakh's cold arid landscape.

Pastoralism is one of the major traditional occupations in Ladakh. It plays a significant role in sustaining the livelihoods of a large segment of the population. People primarily graze animals such as yak, goat, lamb, and sheep. In addition, horses and dogs are also reared. Pastoralism is mainly practiced by communities living in the highlands, many of whom are nomadic and follow seasonal migration patterns. Several scholars have conducted studies on pastoralism in the Ladakh region, including Ladon et al. (2023), Akand et al. (2017), Raj et al. (2020), Khan (2024), and Loat et al. (2024). Their findings indicate that pastoralism has a substantial impact on the economy and income of the people of Ladakh.

2.4 Tourism

Tourism in the Ladakh region is one of the key sectors driving economic development and is a major source of livelihood for the local population. The panoramic landscape provides a suitable foundation for tourism development (Dolma and Kumar 2025; Dar et al. 2022). The impact of tourism on community well-being and sustainable development is significant, as it generates income and employment opportunities for the educated youth of the region (Kapoor and Wangdus 2022). Eco-tourism is an emerging form of tourism in Ladakh. Community participation in the development of eco-tourism can help generate income and support the livelihoods of economically disadvantaged people (Wania et al. 2025; Jaiswal 2024). At the same time, mass tourism is a notable feature of the region. A large number of tourists visit monasteries and scenic natural locations (Lundup 2020).

Many scholars have conducted studies on tourism development and its impact on the economy and rural livelihoods in Ladakh. Several factors influence tourist arrivals in the region. Dar et al. and Pelliciardi (2021) conducted a study analyzing the factors affecting both international and domestic tourist arrivals from 1974 to 2020. Similarly, Dar et al. assessed the carrying capacity of tourism in the Ladakh region and observed that the carrying capacity in major tourist destinations is limited. An earlier study on the dynamics of tourism development and sustainability in Ladakh was conducted by Sood (2000), in which he established a relationship between the evolving nature of tourism and its role in promoting sustainable development.

2.5 Population

The population of Ladakh was 2.74 lakh in 2011, compared to 2.37 lakh in 2001 (COI 2011). It accounts for 0.02% of the Indian population. Leh is the largest district in the country in terms of area. It has 112 inhabited villages and one uninhabited village. Buddhists constitute 77.30% of the population, Muslims 13.78%, and Hindus only 8.16%.The population in Ladakh is sparsely distributed. Most people live in the valley regions along the rivers, as the valleys are expansive and water is available. The other parts of Ladakh are barren and stony.

Leh is the capital city of Ladakh, located in the Trans-Himalaya along the northern tributary of the Indus River. Agricultural land use had been the economic mainstay in the region (Dame and Nusser 2008). Over the last few decades, the urban agglomeration has become attractive for migrants from within and outside Ladakh. Development of tourism also led to urbanization in Leh and Kargil cities (Juliane et al. 2019). Increasing tourists and urbanization has led to water scarcity, which the region is facing.

2.6 Changing Climate and Its Impact

Many scholars have studied the changing climate of Ladakh and its impact on daily life and economic activities such as agriculture, livestock farming, and groundwater availability. Climate change has a significant impact on various aspects of the region's environment, including the declining groundwater levels (Dar et al. 2024), changes in seasonal snow patterns (Passang 2024; Mandal et al. 2024), temporal variations in water quality (Ali et al. 2024), rainfall variability (Singh and Bhatla 2025), and the interlinkages between climate change, water resources, and agriculture (Tuladhar et al. 2023).

2.7 An Unforgotten Journey to the Land of Lamas

In the summer of 2022, I traveled to Ladakh. It was a long journey that started from Aizawl, in northeast India, via Delhi to Leh. The journey was completed entirely by air. It was a trip from the extreme northeast to the extreme northwest, transitioning from a tropical humid climate to a cold climate. The flight from Aizawl to Delhi is 2 h and 30 min long, but including all other formalities, it takes about 5 h and 50 min. After staying in Mahipalpur for a night, the next day at 4:20 am, I boarded an Indigo aircraft. The flight took approximately 1 h and 20 min, and it reached Leh early in the morning at 5:40 am. The air route was spectacular. Our flight passed over Chandigarh and Dehradun before flying over the mighty Himalaya. The Himalaya was covered by dense clouds, with a few of the highest snow-clad peaks visible. Beyond the Himalaya, the territory of Ladakh began. The sky was clear and deep blue. The gigantic, barren, arid mountains, topped with snow, created a panoramic landscape. In the river valleys, patches of green vegetation along with human settlements were visible. The population distribution was sparse, as the total population of Ladakh was about 2.74 lakh. The city of Leh is situated near the Sindhu River, one of the longest rivers in India.

Leh, a tiny city, is situated at an average 3400 m altitude (Fig. 2.2). It is characterized by hot and dry summer and cold and wet winter. Four months of the winter,

Fig. 2.2 The Leh City, surrounded by the mighty Himalayan mountains and the Sindhu River valley. Photo By author

it remains snow-clad. Due to its latitudinal extension, the sun rises at 5:30 am and sunsets at 7:15 pm during the summer, mainly in July. The sun strokes are very intensive during the summer, which leads to sun burning and other health problems. Tourists need sunglasses, sun creams, and caps. The city is surrounded by gigantic barren mountains and snow-capped mountains of Trans-Himalaya, below which is the mighty Sindh River. Its surrounding holds lush green vegetation and arable land. The city center is spectacular, a flat area and it's both sides are shops, which are selling varieties of goods, cloths, fruits and vegetables. To reach the city center, it is slope to climb from all sides. The architecture of the buildings in the city is traditional, mainly made of wood and stone. Varieties of food and beverages are served in the hotels. Owing to high altitude, respiration problem and headache is common therefore one day of rest is essential after reaching Leh. However, Leh is the most accessible and a trade town of Ladakh, which lies 260 km from Srinagar (east). We followed the instructions of the local government and took a rest for a complete day. At the mid-day, it was very hot during the summer whereas at the evening, it was pleasant. Tourist across the country and abroad visit Leh city during the summer. Leh city is located in the Leh district. The district has 113 villages and a total of 1.17 Lakh populations as per the Census of India 2011.

As per the tour package, the second day was devoted to sightseeing in the surroundings of Leh city. It was about 5 hours of sightseeing, starting from the Army Camp, known as the Hall of Fame. It is a Museum with various sections with an important section is Vijay Gallery, where various kinds of weapons used during the Kargil war, are kept. Wall lies inside which is named as "Lest We Forget." Kargil

Fig. 2.3 Spituk Monastery, located 8 km from Leh city on the hilltop facing to the city, behind which is the Sindhu River with a stretch valley, where vegetables are grown largely. Photo By author

war is known as "Operation Vijay" and a documentary about it was shown during the visit of Museum.

After a short visit of the Hall of Fame, we proceeded to Spituk Monastery, situated on the hilltop below which lies the stretch valley of the Sindhu River (Fig. 2.3). Spituk Monastery was founded by Lama Od-de under the Red Hat sect in the eleventh century. Later on, Yellow Hat sect taken over this monastery in the fifteenth century. This is the place known as the rising of an exemplary religious community, visited by Lotsewa Rinchen Zangpo. Spituk means exemplary in Tibetan. A festival renowned as Spituk Gustor is celebrated on 27th to 29th days of the eleventh month of Tibetan calendar mostly falls in January. A statue of Vajrabhairva is unveiled on the day and later on kept in the Mahakal Temple, located on the top of the Monastery. It has a temple of Tara Lhakhang where a statue of the female Bodhisattva Tara lies. It houses about 100 monks with three branches of Monastery—Stok, Sankar, and Saboo. They celebrate Gestor at Spituk together.

After visiting the Spituk Monastery, we proceeded to the Sangam of the Zanskar and Sindhu Rivers near Nimmu village (Fig. 2.4). On the way, we passed through Leh State University, which was established after 2015. The university campus is small, and the landscape is barren and stony. Along the way, a Sikh Gurudwara, run by the Indian Army, is located. The Sikh Gurudwara is famous for serving food (Langar) to every visitor free of cost. It is followed by the Magnetic Hills. Every tourist stops at the famous Magnetic Hills to witness its miracle. The land appears sloped; however, due to the magnetic effect, objects remain stable on the ground.

Fig. 2.4 Zanskar River meeting with the Sindhu River near Nimmu village from the left bank. Photo By author

The Sindhu and Zanskar rivers create a spectacular landscape at the Sangam. It was mid-day, very hot, and the weather was intolerable. The Zanskar River joins the Sindhu River from the left side. Since it was summer, with high temperatures, snow melting was significant, and as a result, the rivers were muddy. The high volume and velocity of the rivers caused tremendous erosion on both sides of the riverbanks. Besides some patches of green vegetation in and around the villages, the entire area was stony, barren, and deserted. During winter, the water in both rivers freezes. The entire route of about 50 km was panoramic. Young tourists often prefer bike riding to travel to the Sangam. We returned to our hotel in Leh at 4:00 pm.

The Sindhu River is one of the longest rivers (3180 km) of South Asia. It originates from the Tibetan High Plateau near Mansarover Lake. It flows northwest through the Ladakh region of India, enters Pakistan from the Gilgit-Baltistan region, turns to southward through the plains of Pakistan before entering into the Arabian Sea near Karachi. It has many tributaries including Jhelum, Chenab, Rabi, Beas, and Sutlej rivers (eastern tributaries); Kabul, Swat, Kurram, and Gomal rivers (western tributaries) form 1.12 million km^2 area. The Zanskar River is a north flowing tributary has three main tributaries—Doda, with source near Pansi-La Pass, the Kargyag River, originates near Shingo La, and the Tsarap River, originates near Baralacha La. The Zanskar River forms a gorge and finally joins the Sindhu River near Nimmu of Ladakh region. Zanskar valley is known for trekking and river rafting and it is often called as "Chadar Trek."

Fig. 2.5 A snowy mountain located before Khardung Pass, on the way to Nubra. Photo By author

2.8 From Leh to Nubra

On the third day, we traveled long from Leh to Nubra, about 140 km. From Leh to Khardung Pass (also known as Khardongla Pass), the road goes through straight upward steep slope about 40 km, makes many curves. On the top, a snowy mountain and below a long and stretch valley of the Sindhu River makes spectacular landscape (Fig. 2.5). These snowy mountains are the source of many perennial streams and rivers. After a short distance, Khardung Pass is located at 5400 m, traversed by the world's highest road. It is the water parting of the two rivers—the Sindhu and Shyok. In this area, Yaks are grazed during the summer (Fig. 2.6). Since, it is the highest place, oxygen was very scarce and some tourists were facing breathing problem. At the time of proceeding from Leh, we bought two oxygen calendars as a safety measures. The winds were very sharp and cold. There were some tea shops serving tea for the tourist. After Khardung, the road goes to downwards in the Shyok River watershed. A small river originates nearby the Khardung pass (glacial-fed) flows along the road and makes some oasis along its course.

After flowing around 30 km, it makes a stretch valley landscape, which is lush green with the pleasant climate. The place is called North Pulu. Tourists stay there for a while and enjoy the landscape, as it gives relief from the hot sun stroke. Tea and snacks are served by the local traders. After North Pulu, the climate becomes hot and the condition of road remains worst. It further goes down along the giant stony and barren mountains without any forest and grassland. And at the end, we reached to the stretch valley of the Shyok River and turned to its downstream areas.

Fig. 2.6 Yaks are grazing near Khardung Pass (5400 m), the highest road of India goes through the pass. Photo By author

The road goes up to the Nubra valley through deserted areas, making sand dunes arid topography.

It was very hot noon, and weather conditions were intolerable. The landscape was all stony and deserted, making arid landscapes. After continuing gone down, we reached the Shyok River valley. Road bifurcates—one goes to Nubra Valley through its downstream course and along the left bank and other goes to Pangong through its upstream crossing the Shyok River several times. On the way, a beautiful monastery is situated in a village called Diskit (3300 m) or Diskit Gompa, belonging to the fourteenth century, founded by Changzen Tserab Zangpo, and believed to be the oldest one in this valley (Fig. 2.7). This monastery is located on a big rock, surrounded by desert and stony landscape. Atop it is located the huge statue of Maitreya Buddha. A famous festival Dosmoche is celebrated here in December every year. This festival is renowned as "Festival of the Scapegoat." During the festival, the mask dance also known as cham dance is performed. We had a short stay in Diskit, first visited the Monastery. The Buddhist philosophy left a strong footprint in my mind and on the several occasions, I enjoy listening the preaching of Budha and further discourse on the Buddha philosophy. And therefore, I had a very fruitful discussion with Bodh monks about the Buddhism at the Monastery. Before proceeding to Nubra service center, we had a delicious lunch—Rajma curry, rice, and potato curry. The owner of the restaurant was from the Himachal Pradesh.

Diskit is the administrative center of the Nubra Valley. It holds a rich biodiversity. Due to the availability of water, it makes oasis and apricot, potato, green leafy vegetables, tomato, cauliflowers, wheat, and mustard are grown here. Among

Fig. 2.7 Diskit Monastery at Diskit, on the way to Nubra Valley. Photo By author

animals, yaks, goats, and cattle are reared. Three important natural features can be seen in this village, are dunes, agricultural patches, and the wide and stretched Shyok River valley (Fig. 2.8).

The Nubra valley is located in the Leh district between Karakoram and Ladakh ranges. The valley is renowned for camel safari, river rafting, trekking, and sunbath. Large sand dunes are found at Hunder, close to Nubra service center. Nubra valley is renowned as Ldurma or the valley of orchard/flowers. It is a broad valley with lofty mountains on its all sides. The Nubra Valley, once on the trading route that connected eastern Tibet with Turkistan via the famous Karakoram Pass, is the most recently opened area of Ladakh. Nubra invites you to a trip back in time into a nearly forgotten "Shangri La." With breathtaking views and delightfully hospitable inhabitants, it is quite an experience. We stayed for an hour in the monastery and further moved toward Nubra, situated along the stretch valley of the Shyok River. Between Diskit village and Nubra, there are many dunes formed (Fig. 2.9). A small population of Bactrian camel exists in the Nubra valley of Ladakh. They are double humped—plump and pliable. During the late winter when pastures are scanty, the humps collapse. Skull bone is comparatively shorter and wider than dromedary camel.

Nubra is a service center and a tourist place, located in the stretch valley of the Nubra River, A tributary of the Shyok River (Fig. 2.10). Every tourist, who visits Ladakh, also visits Nubra. The route makes triangle—Leh-Nubra, Nubra-Pangong, and Pangong-Leh. Winter is frozen and during the summer, daytime is very hot and morning and evening is mild climate. Many tourist companies have camps and lodges in Nubra, where tourist stays for a night. We reached the Cozy ethnic camp

Fig. 2.8 Nubra Valley, Shyok River is flowing along the base of mountain. Photo By author

Fig. 2.9 (**a**) Sand dunes, close to the Nubra service center. (**b**) Camel Safari, double humped camel. Photo By author

at 2:30 pm and stayed in the tents overnight. It was a hot day, so could not go out for sightseeing immediately. Close to Nubra, a camel safari was organized after 4:00 pm along the course of the Shyok River. It was a place where large sand dunes are found. Along the dunes, a stream flows, which is a branch of the Shyok River. After 4:00 pm, the temperature slows down and tourists go for Camel Safaris and for a walk. We went to these dunes after a walk of about 3 km and came back. Dinner was tasty with rice, roti, pulses curry, and mixed vegetables. Curd was also served. The night was very pleasant, as temperature slows down. Nubra valley grows large scale apricot and this was the season for ripening of apricot and therefore, we also enjoyed eating apricot. The climate of Nubra valley is typical, too hot in the summer and too

Fig. 2.10 Stretch Nubra Valley, deserted, and greenery can be seen in a few areas. Photo By author

cold in the winter and day and night temperature variation is high. Therefore, it forms dunes, along with green vegetation patches.

2.9 Nubra to Pangong

After a night stay in Nubra, we left for the Pangong, a highland lake, situated at an altitude of 4300 m. It was an approximately 150 km long road; gig-gag and undulating that took about 7 h to reach the destination. The entire route was stony barren mixed-up with sands, pebbles, and boulders. Several glacial-fed perennial streams can be seen on the way. Wherever, these streams flow, green vegetation patches with sparse settlements are situated. About 70% of the road traverses the Shyok River valley, which has various topographies and climates. The entire valley is very hot mainly in July and August. Water is abundant because of the Shyok River, which is a glacial-fed perennial river with high volume and velocity during the summer and the winter; it is frozen above 3000 m altitude.

The Shyok River, a glacier-fed, originates from the Rimo glacier, one of the branches of the Siachen glacier, flows through the northern Ladakh region and inlets in the Sindhu River in Gilgit-Baltistan, in Pakistan. It has a total of 550 km river course. It forms the beautiful valleys—Shyok and Nubra. Before Nubra valley, the Nubra River joins the Shyok River from the right bank. The Nubra River originates from the Siachen glacier in Ladakh.

The road goes back to Pangong about 80 km along the Shyok River. It bifurcates on the midway, one goes to Khardung Pass and then Leh, and another goes to Pangong. Now the toughest journey starts. First, we traveled along the left side of the Shyok River toward its upstream and then crossed an iron bridge and moved through its right bank. The water of the Shyok River was muddy and reddish and the volume and velocity was high. Between 11:00 am and 2:00 pm, its volume and velocity increases because of snow melting. Often it is flooded and its water blocks the road. The vehicles do not move until the flood water comes down. The road goes up to 40 km and then turns to its another tributary, which is known as the Shyok Gang River (Fig. 2.11). The Shyok Gang River originates from the Pangong highland lake and meets with the Shyok River at Shyok village (3400 m) where an Indian Army base camp is situated. A road bifurcates to the Galwan valley; a transboundary valley that divides India and China. The Galwan valley came into light in 2020–2021, when there was a clash between the Indian and Chinese Armies and many soldiers of both armies died. The entire route goes above 3200 m. The other road goes to Pangong Lake and covers about 20% of areas along the Shyok Gang River. On the way, there are narrow and stony valleys, river gorges, waterfalls, and

Fig. 2.11 (**a**) The Shyok River, muddy with high volume and velocity. (**b**) The Shyok Gang River joining the Shyok River at Shyok village, it further leads to the Galwan Valley. (**c**) The Shyok Gang River flowing through the deep valley. (**d**) A small stream joining the Shyok Gang River on the way to Pangong highland lake. Photo By author

river valley oasis where flowers were blossoming. We stayed at the beautiful bank of the Shyok Ganga River and slowly moved upwards. The entire route was beautiful. The water of the Shyok Gang River was crystal clear and fresh and it was making a spectacular river valley landscape.

Furthermore, moving upwards, another glacier-fed stream joins the Shyok Gang River and makes spectacular river valley landscape. We followed the course of the stream and continued journey up to Durbuk village, a service center and an army camp. On the way, two villages—Agham and Shyok are located. From here, one road goes back to Leh through Changla Pass and other goes to Pangong. We had a lunch at Durbuk village. The road goes through a tableland and also along the valley of a stream. On the way, alpine pasturelands, where horses, goats, sheep, lamb, yaks, and cattle are reared, were visualized. In some other places, we saw a few endangered faunal species (Fig. 2.12). Yaks are the backbone of Ladakhi economy mainly in the highland. They provide food—milk and meat to the people. They also used for transportation of goods. Horses are reared in very high-altitude areas, the major means of transportation. Goat, sheep, and lambs have multiple uses for sustaining livelihood. They are used for milk, meat, and wool. Ladakh is a cold desert. During winter, the entire region is frozen. Wools from goat, sheep, and lamb are used for warm woolen clothes and these clothes are very typical. Many freshwater glacial streams join the Shyok Gang River and enhance its volume and velocity. We left the valley of the Shyok Gang River and proceed to Pangong Lake. The road goes through the last and beautiful village named Tangtse. In surrounding of the village, wheat crop was ripening and was ready to harvest. Some vegetables were

Fig. 2.12 Alpine species of wild animal on the way to Pangong. Photo By author

growing. The settlements were made of wood and stone and rural landscape was spectacular.

After traveling about 2 h from Durbuk village, we reached Pangong Lake, also known as Ludung, at 4:00 pm. Although, it is a highland lake with at about 4300 m altitude, the weather was very hot and dry. The temperature accedes to 27 °C and at night, it recedes to about 5 °C in August. There were many temporary huts and tents for a night halt. The high variability in weather conditions made it difficult to acclimatize to it. While leaving Leh, our travel agent advised us to buy oxygen cylinders at least two because of lacking oxygen. We bought the cylinders, thinking that Pangong Lake is situated at a high altitude and no trees are growing therefore, there may be an oxygen problem occurred. However, the situation was different. We did not feel lacking oxygen. We stayed in the tent, took rest, and completed some pending academic works; and around 6 pm, we went to the bank of the lake, had some photographs, and walked about 4 km both ways along the bank of the lake. After which a dinner and went to bed.

The Pangong Lake is approximately about 220 km long, of which, 65 km (30%) of the lake falls in the Indian Territory (Fig. 2.13). The remaining lake area is occupied by China. This is a trans-boundary lake between India and China having Line of Actual Control (LAC). This region is also known as Aksai Chin, which was a part of Indian territory and China occupied it after the Sino-India war of 1962. Now, the armies of both countries have deployed on both sides of the lake. We stayed at Sangam Cottage for a night. In Pangong, there was no electricity. The cottage owner

Fig. 2.13 Panoramic view of Pangong Lake surrounded by baren landmasses and snow cover mountains. The lake is glacier fed and perennial. It changes its colour several times in a day and makes a beautiful view all the time. Photo By author

arranged electricity by diesel generator only between 7:00 pm and 10: pm (3 hours). No internet connectivity was available. We got up early in the morning and walked along the bank of Pangong Lake.

2.10 Pangong to Leh

After having breakfast, we moved back to Durbuk village, where road bifurcates to Leh via Changla Pass. We followed the same path reached Durbuk via Tangtse village. It was a total of 40 km long road. Another road goes to Nubra via Shyok village. After traveling about 30 km, we reached a beautiful place called Tsoltak. A tributary of the Shyok Gang River makes a beautiful valley where varieties of flowers blossom. It is about a 5 km long valley that ends near Changla Pass. It originates from a huge glacier, close to the Changla Pass (Fig. 2.14). We stayed here along with several tourists for a while and enjoyed ourselves. Changla Pass is located at 5200 m altitude, water parting between the Sindhu and Shyok Rivers. It has a unique landscape and it known as world's highest place connected by road. Road goes downwards from this place. Because of its height, a few tourists face respiration problems. For a moment, we stayed at Changla Pass and had snacks. The tourists were busy in making reals and taking photographs and the road on the top of the Trans-Himalaya was crowded.

The road condition was bad, throughout the way from Pangong to Changla and further. It was broken in many places, and dusty and stony. After traveling a long distance, Shakti village was situated just a few kilometers before Leh city (Fig. 2.15). The village has beautiful rural landscape including settlements and a flat arable land. A small perennial stream, originating from the Changla Pass and draining in the Sindhu River basin, making oasis in the Shakti village and providing ample water for drinking and irrigation to the people inhabiting here. The village has very

(a) (b)

Fig. 2.14 (**a**) A huge glacier making "U" shaped valley close to Changla Pass. (**b**) A beautiful perennial stream, a tributary of the Shyok Gang River, making a spectacular river valley landscape at Tsoltak, originates from the glacier. Photo By author

Fig. 2.15 Shakti village lays upslope of Leh town. A vast wheat crop fields can be seen within settlements. Photo By author

ideal location. It has an Army camp and a beautiful monastery on the other side of the village, on a hill slope.

This perennial stream makes a wide and stretch valley with availability of the vast arable land and evergreen vegetation. People of the village grow summer wheat which is harvested mainly in the August and September. Besides, vegetables are grown here mainly green leaves and seasonal cauliflowers. Among the fruits, apricot and walnut are famous crops with substantial production. Mainly Buddhists inhabit in this village. A few Muslim families are also inhabiting. Moving further to Leh, we passed through Kharu, Satkna, and Thinkse settlements of the Sindhu River basin. It was 2:00 pm when we reached Leh and once again at evening visited the city center of Leh.

The next day, early in the morning, we traveled to Kargil through the Zanskar Valley. It was a marvelous journey. The route was full of spectacular landscapes, including river gorges, terraces, V-shaped valleys, and sandy and stony mountains. In a few areas with stretches of valleys and available water, oases full of greenery were situated. Kargil town is beautifully located on the banks of the Kargil River, creating a spectacular cityscape. A campus of Ladakh University is located here, where students from across the Ladakh Union Territory pursue their studies. After visiting a few places in Kargil, we returned to Leh late at night.

Kargil is among the two districts of Ladakh with its headquarters in Kargil service center, situated on the bank of the Kargil River (Fig. 2.16). The district has a population of 1.25 lakh as per the Census of India 2011. Population is scattered over an area of 14,086 km^2. It is located 205 km from Srinagar and about 234 km from

Fig. 2.16 Kargil urban center, headquarters of Kargil district, situated on the banks of the Kargil River. Photo By author

Leh. Kargil is called the land of Agas and it is inhabited mainly by Shia Muslim and Agas. They are religious heads and preachers. Kargil means castle center, a place between many kingdoms ruled. It also means people can stay anywhere. Kargil is located equidistant from Srinagar, Skardo, Leh, and Padum. A book on Kargil Blunter is written by historian Parvez Dewan, which supports this fact. Kargil history is long and rich and it was ruled by many rulers. From May to July 1999, Kargil war was fought between India and Pakistan and India won the war after a huge loss of soldier life from both sides. On 31 Oct 2019, after abolition of article 370, Kargil came under the Union Territory of Ladakh. The ancient Kargil was known as Purik. The name was given by the Tibetan scholar. Drass area is inhabited by Dard race people and Zanskar has Ladakhi—Tibeto stock. Aryans and Mongoloids are other races, forming the place multi-ethnic, multi-linguist, and multi-culture. Brogpas, Baltis, Purik, Shinas, and Ladakhi are the major types of people inhabiting Kargil and these are the major languages spoken. Balti and Shina languages are written in Urdu script and Urdu is common language. In the fifteenth century, Islam came to Kargil and slowly, the people follow Islam.

2.11 In Surroundings of Leh

The next day, a tour was organized around Leh. The major destinations visited were Thiksey Monastery, Hemis Monastery, Sindhu Ghat, Shey Palace, and the 3 Idiots Fame Rancho School. Most parts of the film *3 Idiots* were shot in Ladakh, including at Pangong Lake. The school, named Rancho School, is located close to Leh city and has now become a famous tourist attraction in Ladakh. This part of Leh is beautiful and lush green, with many world-famous monasteries. On that day, the Dalai Lama was visiting Leh and had a preaching program for his devotees (Fig. 2.17).

Fig. 2.17 Dalai Lama, Buddhist Monk preaching at Shey Palace, close to Leh. Photo By author

The Dalai Lama holds a significant place in the Buddhist community. He is a religious guru and the 14th in the lineage of Dalai Lamas, considered the supreme religious leader of Buddhist people. In 1956, he was exiled from Tibet after China conquered it. The Dalai Lama fled to India and was given shelter in Dharamshala, Himachal Pradesh, where he currently resides. Leh is his second home, and he frequently visits the city on various occasions. Due to his presence in Leh that day, there was a huge crowd of devotees, making it difficult to move around, and it took a lot of time to cover all the tourist destinations.

One of the important destinations was Sindhu Darshan. The Sindhu is a Sanskrit word, sacred to Hindus and it is mentioned in the Rigveda, the oldest Hindu scripture. Its English name is Indus and Persian word is Hindu. Sindhu is personalized as an emperor and believed to be the son of Lord Vishnu. Therefore, the followers of Buddhism and Hinduism are worship it. On Guru Purnima, a festival named "Sindhu Festival" is organized in the banks of the Sindhu River, every year. The festival includes a reception ceremony, cultural programs, a sightseeing tour, and a bonfire. The world-famous Indus Valley Civilization (3000 BC) was originated and flourished in the Sindhu River valley; therefore, the Sindhu River plays an important role in shaping the ancient culture. The civilization is known as the oldest and primitive. It was the great example of engineering masterpiece. It was the center of human civilization and fishing that enhanced the livelihood of nomadic fishing communities, the remnants are still visualized today. Furthermore, it was the center of trade and cultural exchange between Indian subcontinent with Central Asia and the Middle East. In due course of time, many factors including geography and climate change resulted in the end of the great civilization.

2.12 Conclusions

This chapter "In the Land of Lamas" reveals about an extensive visit of Ladakh and a comprehensive study of its land and people. The study includes the natural, cultural, social, and economic settings of Ladakh. The spectacular landscapes—the mountains, river valleys, agriculture, and forests make the Ladakh unique and diversified, which promote for sustainable tourism development. Ladakhis have rich cultural heritage and the people are innocent. The land is characterized by the high mountains and the largest monasteries. It is a land of Buddha. This chapter details about the different popular tourist routes—Leh to Nubra, Nubra to Pangong, and Pangong to Leh, and a visit of the surrounding areas of Leh. The climate, agriculture, forests, and rivers of Ladakh are described in detail.

References

Akand AH, Singh BP, Chander M (2017) Contribution of livestock in livelihood: Ladakh region. Indian J Anim Sci 87(5):649–652. https://doi.org/10.56093/ijans.v87i5.70273

Ali G, Chaudhari M, Shah P, Shrivastav P (2024) Temporal changes in water quality in Leh Ladakh region: impact of urbanization. Environ Res Technol 7(4):637–664. https://doi.org/10.35208/ert.1431710

COI (2011) Population of Ladakh. Registrar General of India, New Delhi

Dame J, Nusser M (2008) Development perspectives in Ladakh, India. Geogr Rundschau Int Ed 4(4):20–27

Dar SN, Shah SA, Wani MA (2022) Geospatial tourist information system for promoting tourism in trans-Himalayas: a study of Leh Ladakh, India. GeoJournal 87:3249–3263. https://doi.org/10.1007/s10708-021-10431-4

Dar FA, Ramanathan AL, Mir RA, Pir RA (2024) Groundwater scenario under climate change and anthropogenic stress in Ladakh Himalaya, India. J Water Climate Change 15(4):1459–1489. https://doi.org/10.2166/wcc.2024.307

Dar SN, Shah SA, Wani MA (2025) Tourism carrying capacity assessment for Leh town of Ladakh region in Jammu and Kashmir. International Journal of Current Research. https://www.journalcra.com/article/tourism-carrying-capacity-assessment-leh-town-ladakh-region-jammu-and-kashmir

Dolkar R, Mehdi M (2025) Low-cost sustainable technology for winter vegetable cultivation in lower belt of UT Ladakh. Int J Agric Extens Soc Dev 8(2):332–336

Dolma S, Kumar A (2025) Ecological resilience and economic shifts: tourism and solid waste in Ladakh's mountainous landscape. Asian Res J Arts Soc Sci 23(3):92–103. https://doi.org/10.9734/arjass/2025/v23i3653

Down To Earth. (2025). Greenhouse innovations are giving new life to farming in Ladakh. https://www.downtoearth.org.in/agriculture/greenhouse-innovations-are-giving-new-life-to-farming-in-ladakh

Hayward GW (1870) Journey from Leh to Yarkand and Kashgar, and exploration of the sources of the Yarkand River. J R Geogr Soc Lond 40:33–37

Indian Council of Agricultural Research (ICAR) (2024). Grape cultivation expands in Ladakh with new initiative. https://icar.gov.in/grape-expansion-high-altitude-region-ladakh-and-leh

Jaiswal I (2024) Assessing tourism sustainability: a case study of Ladakh region. In: Singh P et al (eds) The need for sustainable tourism: regional perspectives. Emerald Publishing, Leeds, pp 151–172. https://doi.org/10.1108/978-1-83608-668-020241023

Jina PS (1996) Ladakh: the land and the people. Indus Publishing, NewDelhi

Juliane D, Susanne S, Judith M, Marcus N (2019) Urbanisation and socio-ecological challenges in high mountain towns: insights from Leh (Ladakh), India. Landsc Urban Plan 189(2019):189–199. https://doi.org/10.1016/j.landurbplan.2019.04.017

Kale VS (2014) Landscapes and landforms of India. Springer, Cham

Kapoor S, Wangdus J (2022) Impact of overtourism on sustainable development and local community wellbeing in the Himalayan region. Int J Leisure Tour Market 8(4):199–214. https://doi.org/10.1504/IJLTM.2022.122454

Kashmir (2006) Encyclopedia Americana. Scholastic Library Publishing, New York

Khan AA (2024) Feeding practices in rearing of yak and its bovine crosses in Ladakh. Indian J Anim Prod Manage 40(3):211–214. https://doi.org/10.48165/ijapm.2024.40.3.13

Ladakh Autonomous Hill Development Council (LAHDC). (2024). Demonstration farm for medicinal and aromatic plants in Ladakh. CSIR-Indian Institute of Integrative Medicine. https://iiim.res.in/csir-iiim-lahdc-to-set-up-demonstration-farm-for-maps

Ladon P, Nüsser M, Garkoti SC (2023) Mountain agropastoralism: traditional practices, institutions and pressures in the Indian trans-Himalaya of Ladakh. Pastoralism 13(30). https://doi.org/10.1186/s13570-023-00289-1

Lamo K, Stanzin D, Sheikh FD, Lhaskit J, Angchuk S, Safal R (2024) Exploring the potential of tomatoes in Ladakh's cold arid landscape through passive solar greenhouse cultivation. Int J Agric Extens Soc Dev 7(2):54–58. https://doi.org/10.33545/26180723.2024.v7.i2Sa.333

Loat S, Pukhrambam M, Mepfhuo K, Medhi D, Paul V, Hussain M, Jatana MS, Bhatt N, Sarkar M (2024) Yak—the "almighty livestock" of the rugged Himalayan Mountains. J Livestock Biodivers 13(1):157988. https://ebook.icar.gov.in/index.php/JLB/article/view/157988

Lundup T (2020) The sacred and the secular: exploring mass tourism encounters at Lamayuru monastery in Ladakh. Himalaya 39(2):119–130. https://doi.org/10.2218/himalaya.2019.7856

Mandal A, Vishwakarma BD, Angchuk T, Azam MF, Garg PK, Soheb M (2024) Glacier mass balance and its climatic and non-climatic drivers in the Ladakh region during 2000–2021 from remote sensing data. J Glaciol. https://doi.org/10.1017/jog.2024

Mann R S (2002) Ladakh then and now: cultural, ecological, and political. Mittal Publications New Delhi p. 15

Passang S (2024) Comparative perspectives on the seasonal snow patterns in Ladakh. In: Snow cover distribution and dynamics. Springer, Cham. https://doi.org/10.1007/978-3-031-57692-8_4

Pelliciardi V (2021) Factors affecting international and national tourist arrivals (1974–2020) in Leh District. Eur J Sustain Dev 10(1):736. https://doi.org/10.14207/ejsd.2021.v10n1p736

Raj A, Angchuk D, Islam MA (2020) Livelihood dependence on highland pastoralism (doksa) in trans-Himalayan region of Zanskar, Ladakh. Int J Environ Clim Change 10(12):67–76. https://doi.org/10.9734/ijecc/2020/v10i1230285

Reach Ladakh Bulletin (2025) CAZRI-RRS organizes winter exhibition on vegetable cultivation and diversification. https://www.reachladakh.com/news/social-news/cazri-rrs-organises-winter-exhibition-on-vegetable-cultivation-and-diversification

Singh R, Bhatla R (2025) Assessment of rainfall variability in Ladakh amidst evolving climate. Nat Hazards 121(3):3073–3097. https://doi.org/10.1007/s11069-024-06922-5

Sood VK (2000) Dynamics of tourism development and aspects of sustainability: a case study of Ladakh. Tour Recreat Res 25(3):115–127. https://doi.org/10.1080/02508281.2000.11014930

Tewari JC, Pareek K, Tewari P, Raghuvanshi MS (2015) Productivity and energetics of crops in farming systems in Ladakh: a case study of a village in Leh. In Sustaining agricultural productivity in arid ecosystems: challenges and opportunities pp 31–37

Tuladhar S, Hussain A, Baig S, Ali A, Soheb M, Angchuk T, Dimri AP, Shrestha AB (2023) Climate change, water and agriculture linkages in the upper Indus basin: a field study from Gilgit-Baltistan and Leh-Ladakh. Front Sustain Food Syst 6:1012363. https://doi.org/10.3389/fsufs.2022.1012363

Wania MD, Dadab ZA, Bhat WA, Shah SA (2025) Community-driven ecotourism in the trans Himalayan region: a sustainable model for cultural and environmental preservation. Community Dev. https://doi.org/10.1080/15575330.2025.2451358

Chapter 3
Hari Ki Dun: The Valley of Lord Vishnu

Contents

Abstract The Himalaya contains many cultural valleys, each named after local deities. One such valley is Hari Ki Dun, the Valley of Lord Vishnu. It can be reached after a long trek through the Tons River valley. The Tons River is known for its rich cultural heritage and abundant natural resources. The entire valley falls under the Rawain cultural realm, which possesses distinct cultural practices that differ from those of other Himalayan regions, despite the area being economically poor. This

V. P. Sati, *Exploring the Mystical Himalaya*, Springer Geography,
https://doi.org/10.1007/978-3-032-14275-7_3

chapter provides a detailed description of the region's culture, economy, environment, and people. Throughout the year, locals celebrate various fairs and festivals and engage in traditional agriculture, cultivating a variety of fruit and vegetable crops.

Keywords Tons River · Rawain region · Hari Ki Dun · Mahasu Devta · Karna · Polyandry system

3.1 Introduction

"Hari" is considered Vishnu in the Hindu scriptures, and "Dun"means Vihar, a place where Hari is revered. Thus, Hari Ki Dun is also called the valley of God. However, Hari Ki Dun is a natural entity, situated in the lap of the Greater Himalayan snow-capped mountain peaks (Fig. 3.1). Hari Ki Dun is a unique natural feature on Earth, a no man's land above the tree lines and below the snow line. It is a vast alpine pastureland, home to hundreds of faunal and floral species, and numerous rivers, streams, and rivulets that form cascades, waterfalls, and meanders. It is a summer dwelling for animal herders, a land of wandering fairies, witches, and souls, and a place for practicing penance and attaining peace, tranquility, and Moksha (salvation) for sages and saints. The Yamuna, a sister of Yama, the God of Death, along

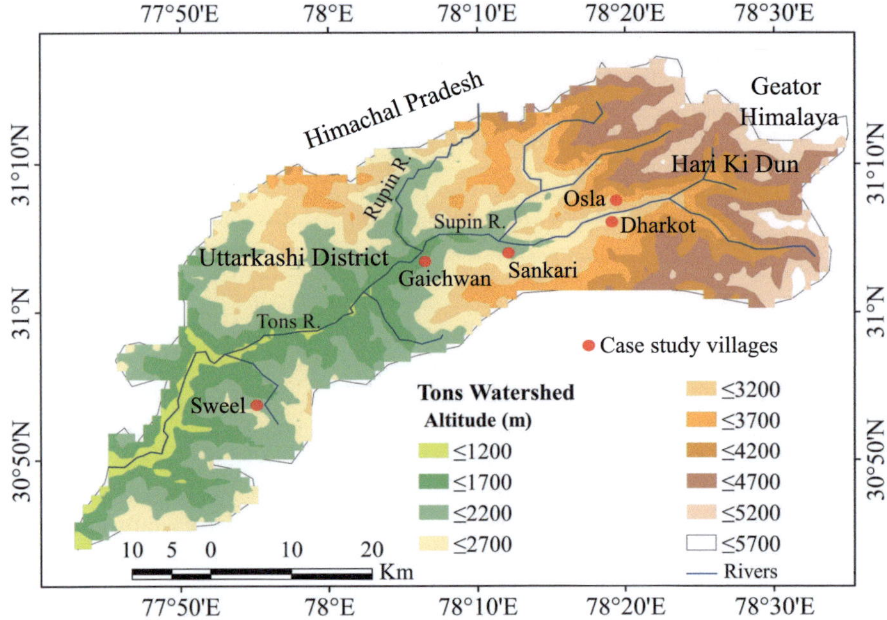

Fig. 3.1 The tons river basin where the famous Hari Ki Doon is situated

with its hundreds of rivers, streams, and rivulets, drains from the Greater Himalaya to the Ganges valley through spectacular landscapes (Singh 2018).

The Himalayan people are innocent, living closely with nature and sustaining their livelihoods from nature's products, such as tuberous fruits. Their occupation is primarily agriculturally based, including farming traditional crops, livestock rearing, and collecting timber and non-timber forest products. The occupation varies according to the altitudes—river valleys, middle altitudes, and highlands. However, it is unique and full of romance. There are many trails that lead to the Hari Ki Dun—Mussoorie-Purola, Kalsi-Chakrata, and from Himachal Pradesh. These trails comprise spectacular natural, forest, and cultural landscapes—river valleys, mountain peaks, forests, and micro-cultural realms.

Agriculture is the main occupation and the primary source of income and economy in the Tons River basin of the Garhwal Himalaya. A wide variety of crop races and cultivars are grown here, reflecting rich biodiversity. However, the traditional methods of cultivation result in comparatively low productivity (Sati and Banerjee 2024). Alongside conventional agriculture, people also practice traditional agroforestry and agri-horticultural systems, which have high potential (Rawat, 2023). Silviculture is commonly practiced in home gardens, and fruits and vegetables are abundantly grown throughout the Tons River basin (Kumar, 2022). In recent years, fruit cultivation—particularly apples—has shown impressive growth across the valley. Livestock farming runs parallel to agriculture and plays a significant role in sustaining rural livelihoods. Kala et al. (2014) noted that the cultivation of medicinal and aromatic plants in the Tons Valley holds great potential for enhancing livelihood opportunities. Additionally, crops such as French beans, peas, and okra have high yields and economic benefits, contributing to the expansion of horticultural farming (Papnai et al., 2017).

The Tons River basin is a remote region with an underdeveloped economy. Even today, several villages lack access to roads and electricity. A large portion of the population depends on agriculture as the primary source of income and livelihood. Although the government has introduced rural development schemes through participatory learning and action, the region has yet to achieve the goals of sustainable rural development (Gangopadhyay & Mukhopadhyay, 2010). Rawat (2018) examined the subsistence economy of the Tons River basin, which is shaped by its fragile ecology and where tourism is gradually emerging as a contributor to the local economy. Tourism is now seen as a promising sector for sustainable development. The valley holds significant potential for the development of homestay tourism, adventure tourism, and cultural tourism (Sati and Banerjee 2024).

3.2 Exploring Hari Ki Dun: Natural and Cultural Aspects

On 3 July 2024, I, along with three professors of higher education from the Garhwal Himalaya, embarked on the Hari Ki Dun trek. We started our journey early in the morning from Dehradun, the capital city of Uttarakhand, situated in the stretch valley of Doon, and traveled by taxi. We followed the Mussoorie-Purola route. After passing through the hill city of Mussoorie, we stopped at Kempty Falls. Mussoorie, a beautiful hill city in Uttarakhand located at an average elevation of about 2000 m and 30 km from Dehradun, is surrounded by stunning temperate forests dominated by deodar trees. Known as the "Queen of the Hills," Mussoorie serves as a summer resort and a haven for tourists. It is well-connected by road to Dehradun. Some of the notable tourist attractions within Mussoorie include Company Garden, Gun Hill, Lal Tibba, Landour, the Tibetan Buddhist Temple, Camel Walk, and Mall Road. The climate during the summer is pleasant, while winter often brings several spells of snowfall. We passed through the heart of Mussoorie and descended to Kempty Falls, located 15 km from Mussoorie on the way to Yamunotri Dham.

The road from Mussoorie to Kempty Falls was narrow, and in several places, landslides had occurred. Since it was the rainy season, the intensity of the landslides was severe, making it very challenging to pass through the affected zones. However, the dense oak forests along the road added to the spectacular beauty of the landscape. Eventually, we reached Kempty Falls, situated at an altitude of about 1340 m in Ram Gaon. Kempty Falls is a stunning site and one of the highest waterfalls in India, cascading from a height of approximately 60 feet and forming several beautiful cascades. After traveling another 10 km, we stopped for breakfast, which consisted of roti and vegetable curry. Traditional Himalayan vegetarian food is mainly organic and incredibly flavourful. However, along the roads leading to popular tourist destinations and pilgrimage sites, food habits have shifted (Sati 2023a). Instead of traditional and nutritious meals, fast food has become more prevalent. We continued our journey and eventually reached the Yamuna Bridge (Mussoorie Bridge), where the road splits: one route leads to Barkot (68 km away), while the other goes to Vikas Nagar and Thatyur. We took the road to Barkot and continued traveling along the bank of the Yamuna River. This route is quite popular as it leads to the Yamunotri highland pilgrimage via Barkot. The road quality was good overall, but there were occasional small and large landslides due to the monsoon season. The monsoon season is a curse for the Himalayan region. The Himalaya, being one of the most vulnerable and fragile landscapes in the world, is highly susceptible to natural hazards and climate change. Both the Yamuna and Tons River basins are particularly prone to geo-hydrological hazards, including landslides, mass movements, flash floods, and debris flows. These events devastate the landscape—damaging agricultural land, settlements, and businesses—and claim many lives each year. Consequently, traveling on these roads was dangerous in many areas.

The Yamuna River, along with its major tributary, the Tons River, forms the largest river system in the western part of the Uttarakhand Himalaya, known as the Yamuna-Tons river system. The other two significant river systems in the Uttarakhand Himalaya are the Bhagirathi-Alaknanda and Kali-Saryu river systems (Sati, 2021). The Yamuna-Tons river system boasts the largest drainage basin in the region, covering an area of 5004 km^2. A massive barrage, 2.4 km in length, has been constructed on the Yamuna River to generate electricity and provide ample irrigation water. This barrage supplies water to two powerhouses: the Dhakrani Power House, with an installed capacity of 84.75 MW, and the Dhalipur Power House, with an installed capacity of 28.25 MW. Together, these powerhouses generate an average annual energy output of 380.51 million units (MU).

Between Yamuna Bridge and Dakpathar (another trail), mangoes and litchis are grown abundantly, and their quality is among the best. The very popular mango variety, Amrapali, known for its pulpy and sweet taste, is cultivated in this area. The climate is hot and humid, and the land is fertile, consisting primarily of plains and river terraces. The altitude ranges between 600 and 800 m. Mangoes are also grown at elevations up to 1000 m along the Yamuna River. However, at higher altitudes, their quality tends to be average, and the yield is comparatively lower.

Along the course of the Yamuna River, numerous homestays can be found, as the road leads to Yamunotri, a renowned highland pilgrimage, and the trekking destinations of Hari Ki Dun and Kedar Kantha. Mailgaon, situated on the bank of the Yamuna River near the Yamuna Bridge, has around 10 homestays. These homestays accommodate tourists throughout the year, especially during the tourist season, with an average of 20–25 visitors per month. However, due to their remote location, far from urban and rural centers, the number of tourists staying in these homestays is relatively low. Nainbag, a town and service center, is also located on the bank of the Yamuna River in the Jaunpur block of Tehri district, near the junction of the Tehri and Uttarkashi districts (Uttar Pradesh District Gazetteers, 2021). This region is ideal for cultivating mangoes, tomatoes, and guavas. After passing through Nainbag, we continued on to the service centers of Damta and Sarai. Approximately 10 km further, we arrived at a small settlement called Kuwa. Here, a small road branches off to Lakhamandal village, located 2 km away, which is famous for its Lord Shiva Temple.

Fig. 3.2 Shiva temple constructed in Nagara style in the twelfth century. Photo By author

All team members wished to visit Lakhamandal to perform a puja for Lord Shiva, so we headed to Lakhamandal village (Fig. 3.2). Lakhamandal is a typical Jaunsari village, where people follow the customs of the Pandavas, and its landscape is truly spectacular. The name "Lakhamandal"is derived from two words: "Lakh," meaning many, and "Mandal," meaning temples. The locals speak Jaunsari and wear distinctive traditional attire. Their houses are traditionally carved, made of wood and stone, and exhibit unique architectural styles. At the heart of the village lies the Lord Shiva temple. As we approached the temple, it began to rain. In the temple premises, priests were seated on a platform made of stone and cement, sheltered by a roof. While we were worshiping the deity, one of the priests approached us and invited us to perform rituals. He narrated the story of the Pandavas and their connection with Lord Shiva and guided us through the temple premises. The priest explained the significance of the various Shiva Lingams in the temple. A red Shiva Lingam symbolizes the Treta Yuga, while a green Shiva Lingam represents the Dwapar Yuga. He also recounted the story of the Lakshyagriha, a palace of wax built by Duryodhana to kill the Pandavas during their Agyatvas (period of exile in disguise). However, the Pandavas, realizing the plot, escaped by digging a cave, which still exists within the temple premises. The Archaeological Survey of India continues to excavate the site, unearthing hundreds of Lingams. We performed the traditional *Jalabhishek* ritual by offering water, milk, and flowers to the Shiva Lingam. The temple, also known as Mahamundeshwar Shivling, holds great historical and spiritual significance.

Lakhamandal village is situated at an average altitude of 1150 m on a gentle slope along the right bank of the Yamuna River. According to the Census of India 2011, the village had a population of 1044 people across 185 families (Gazetteer of India, 1909). Most of the inhabitants belong to Brahmin communities, including the Bahuguna, Nautiyal, and Purohit clans. Despite their Brahmin lineage, they have been classified as Scheduled Tribes by the Government of India. Additionally, the village is home to Scheduled Caste communities. The river terraces surrounding Lakhamandal provide fertile arable land, supporting the cultivation of crops like paddy, wheat, and pulses, as well as fruits such as mango and guava, and a variety of vegetables. These crops are produced with high yields, thanks to the favorable conditions. Every year, on April 15, the village hosts the *Thaura* festival, where thousands of devotees gather to worship Lord Shiva. It is a deeply held belief among the locals that couples without children who worship Shiva throughout the night are blessed with a child. Another major event, the Maha Shivaratri fair, is celebrated in the temple premises, attracting thousands of devotees who come to perform rituals and offer prayers to Lord Shiva.

The Shiva temple in Lakhamandal is an exquisite example of Nagara style architecture, built during the twelfth and thirteenth centuries. The abundance of Shiva Lingams and the unique construction style suggest that the site once housed numerous Shiva temples. Evidence from the stone and brick pyramids found within the temple premises indicates that parts of the temple date back to the fifth and sixth centuries. The intricately carved stone pillars further support the ancient origins of the temple's construction. The present temple is believed to have been constructed by Queen Kumari Ishwara in memory of her late husband, Chandragupta, the son of

King Jalandhara. According to legend, the Pandavas established approximately 0.125 million (125,000) Shiva Lingams in Lakhamandal during their exile (*Agyatvas*). Excavations by the Archaeological Survey of India continue to uncover these Lingams, highlighting the historical and spiritual significance of the site.

We spent a very fruitful time in the temple premises, where many other pilgrims were also visiting and performing *puja*. After concluding our visit, we returned to the main road and proceeded toward Naugaon, approximately 20 km away. Naugaon, situated at an altitude of 1160 m, is a small yet growing town with a population of 2516 as per the 2011 Census. It lies on the route to the Yamunotri pilgrimage and has been expanding into the surrounding arable land due to urban growth. The town serves as a cultural junction, often referred to as *Triveni*, where the traditions of Tehri and Uttarkashi districts converge. The Yamuna River flows gently through Naugaon, dividing the town and adding to its charm. Naugaon is renowned for its agricultural produce, especially its diverse crop cultivars, fruits, and vegetables. Among these, tomato cultivation (Fig. 3.3) stands out as the primary commercial activity, with produce distributed to local and regional markets. The town's strategic location offers two major routes: one leading to Barkot and onward to Yamunotri, and the other heading toward Purola and the Tons Valley. After leaving the main road, which was of better quality, we continued toward Purola on a narrower, less well-maintained road. This marked the beginning of the Rawain sub-cultural region. Up to this point, the landscape had been rough, rugged, and precipitous, showcasing the natural beauty and challenges of the region.

We crossed the Yamuna River and continued toward Purola, with the journey offering breathtaking panoramic views of the surrounding rural settlements. The

Fig. 3.3 A woman selling tomatoes on the roadside in Naugaon. Photo By author

distance between Naugaon and Purola is approximately 20 km, following the Kamal River valley. The Kamal River, a vital tributary of the Yamuna River, merges with the Yamuna at Naugaon. The drive along this scenic route was pleasant, with farmers selling tomatoes in large quantities along the way, reflecting the region's agricultural richness. As we approached Purola, the wide, expansive Kamal River valley unfolded before us, offering a spectacular and picturesque view. The valley is a natural masterpiece and a significant part of the local geography. We took a short halt before reaching Purola to appreciate the beauty of the Kamal River valley. Known as the lifeline of Purola and its neighboring villages, the Kamal River nurtures a fertile and productive valley. This region is particularly famous for its high quality and abundant yield of red rice, a cherished staple in the area. Upstream of the river lies Rama Sarain village, celebrated for its exceptional crop production, especially rice. The fertile lands and ideal climatic conditions make it a hub for agricultural diversity and abundance.

Purola, a beautiful village, is situated on the banks of the Kamal River at an average altitude of 1200 m in the Uttarkashi district of the Garhwal Himalaya (Fig. 3.4). In 2011, its population was 2394, with 586 families. It is a wide and expansive valley surrounded by rich biodiversity, including dense pine forests at lower altitudes and temperate forests at higher altitudes (Jain and Puri 1984), along with a variety of temperate faunal species. Purola is an important center of education, where Rawain culture flourishes. Well-connected by road, it serves as the gateway to the Hari Ki Dun trekking route. This fertile valley of the Kamal River is known for its high production and productivity of red rice and tomatoes.

This region is characterized by three cultural realms: Jaunsar-Bawar, Rawain, and Shiktur Patti (the Western Garhwal region). Jaunsar-Bawar is located in the Tons River catchment, including the valleys of the Rupin and Supin rivers. The Rawain region is situated in the Yamuna Valley, primarily in the Kamal Valley of the Purola Taluk. The upstream areas of the Rupin and Supin rivers are known as Shiktur Patti. The culture and cultural heritage in these areas are rich and diverse.

(a) (b)

Fig. 3.4 (**a**) Stretch valley of the Kamal River. (**b**) Purola town is now spanning. It population has increased many-folds after the Census of India 2011. Photo By author

Fig. 3.5 Traditional food thali: red rice with Rajma, spinach curry, Chatni, pulses soup, fried potato, Kheer, Raita, Manduwe Ki Roti, and Bharwa Puri. Photo By author

After a few kilometers from Purola, we halted at a homestay restaurant and had lunch with traditional food and beverages. Red rice, Manduwe ki Roti, Rajma curry, and Raita were both tasty and nutritious (Fig. 3.5). The homestay is located on the road leading to Mori, in a spectacular landscape at about 1310 m altitude, near Hudoli village. Along the roadside, many apple orchards are situated. This was the season for apple ripening, so the red apples enhanced the environment, making it panoramic and pleasing to the eyes. Below, the beautiful Kamal Valley and Rama Sarain village lay, with dense pine forests between them. The rain had stopped, though it remained cloudy. After a pleasant stay, we proceeded further to Mori, which was about 45 km away. Two rural settlements lie on the way: Bagasi Saani village (1250 m) and Jormara (2000 m). While traveling to Mori, we ascended the hills through dense pine forests and reached Jarmora, a forest camp office. The forest department maintains a large nursery and an apple orchard there (Rana & Datt 1997). Pine trees hold high economic value, and in the Yamuna and Tons basins, pine products are a major source of the economy and wood industry. Pine wood is widely used for making houses and furniture, and the resin extracted from pine is used for making the blacktop for roads and various household items. The Jarmora forest camp area is a watershed for the Tons and Yamuna Rivers. Over a 3-km stretch, the forest landscape is spectacular, making it a potential eco-destination (Sati, 2018). Tourists and pilgrims visiting the Mahasu Devta temple, as well as those trekking to Hari Ki Dun and Kedar Kantha, often stay here for a while. From Jarmora, the road descends into the Tons River valley, passing through a beautiful stream, a tributary of the Tons. Before reaching Mori town, a village is situated on a fertile terrace where many crops are grown.

Mori (1170 m) is located on both sides of the Tons River. It is a town and service center, serving the vast surrounding areas. It is believed to be an ancient settlement, dating back to before Christ, and is influenced by the culture of the Pandavas and Kauravas. Located at the juncture, approximately 30 km from Tyuni and Hanol, and 12 km from Netwad, Mori is well-connected by road. Himachal Pradesh lies on the right bank along the Tons River. Mori town also serves as a trade center and base camp for tourists and trekkers. During the winter season, the upper reaches of Mori receive snowfall, making it a haven for tourists. The town is very popular for serving traditional food, including Manduwe ki Roti, Gahat ke Paranthe, and Rajma curry. The Tons River forms a wide and expansive valley in Mori and its downstream area. Therefore, during the winter season, tourists engage in water sports, such as river rafting. The town is a center of education, with institutions ranging from primary to higher education. Students from remote and surrounding areas come here for their education. It is also a cultural hub, home to temples dedicated to Koteshwar Mahadev, Duryodhana, Someshwar, Karna, and Mahasu Devta. The people of Mori belong to various castes and clans, including Jaunsari, Tibetans, Garhwali, and Rawain.

Netwad, a small village with a population of 319 people from 66 families (2011), is situated on the left bank of the Tons River, about 15 km north of Mori. The village covers a total area of 66.51 hectares. The Rupin and Supin rivers meet at Netwad, and after their confluence, the Tons River is formed. A 30 MW capacity

Fig. 3.6 Typical settlements of Dewera village (1650 m) wooden houses and tin or slate roofs. Photo By author

hydroelectric power plant, constructed on the bank of the Tons River, began producing electricity in November 2023. In the vast upslope areas of Netwad, many villages are located, and these villages now produce high-quality and high-quantity apples. Some of these villages (Fig. 3.6) include Gaichwan (1500 m) and Dewera (1650 m). Netwad serves as the base camp for trekking to Kedar Kantha, located about 16 km by road from Netwad, with a 6-km trek. Kedar Kantha is known for its vast alpine pastureland and summer camps, where animal herders seasonally migrate for grazing, primarily for goats, sheep, lambs, horses, and cattle. Surrounded by the Greater Himalaya, Kedar Kantha is a popular destination for trekkers. The apple growers in the Netwad region are influenced by Himachali immigrants, who have large apple orchards on the upper slopes of Netwad. The people here have a mixed culture, influenced by Jaunsari, Rawain, Garhwali, Himachali, and Tibetan traditions. On the other side of the Tons River lies the Rodu district of Himachal Pradesh, and the northern border is with Tibet, which adds to the cultural influence of Himachali and Tibetan traditions. The farmers grow apples alongside pears, peaches, plums, apricots, and walnuts in abundance, selling apples to Uttar Pradesh. The average income of each apple grower is five million INR per season.

It was a long and tiring journey, about 12 h, from Dehradun to Netwad. Therefore, we decided to stay at a homestay in Gaichwan village, located 5 km upslope from Netwad (Fig. 3.7). The village is situated on a gentle slope, and the landscape is spectacular. The homestay was right on the roadside, and the owner, Mr. Ajit Singh Ranger, is a retired government official. After his retirement, he started growing apples on a 2-hectare plot of land, producing more than INR five million worth of

Fig. 3.7 Homestay at Gaichwan village (1500 m). Beautiful building surrounded by cowshed and apple orchard. Photo By author

Fig. 3.8 An apple orchard is located in Gaichwan village. Photo By author

apples in a season (Fig. 3.8). In addition to apples, he also grows pears, red rice, potatoes, and seasonal vegetables. The homestay is well-equipped with modern facilities, and the rent was nominal. Upon arriving at the homestay, we visited a vast apple orchard. Every family in the village owns cows and a pair of oxen. The animals are kept on the ground floor, called "Obra," while the first floor or above is used for living and storing other items. One room is reserved for storing food grains, and it is called "Kothara."

A traditional and delicious dinner was served to us, comprising Manduwe Ki Roti, Red Rice, Rajma curry, green leaves curry, fresh butter, butter milk, and coffee. Dinner and breakfast was served in the courtyard of the house, which is called as "Tyabari." This is a custom of serving food in the courtyard. The weather condition was pleasant and enjoyable. While discussing about the culture of the region, Mr. Ranger informed us that in Gaichwan village, almost all males of his age group has two wives but he has only a wife and for which he will always repent. The next morning, we visited the Karna Temple in Dewera village. The temple is ancient and constructed in the Memory of Karma, the elder brother of Pandavas. People of this village believe themselves as the descendent of Kauravas mainly Duryodhana. They believe that Duryodhana saved them from adverse situation during his rule in the Tons valley. Netwad and its surrounding villages have four folk deities—Pokhu Devta, Duryodhana, Karna, and Someshwar. However, all temples are dedicated to Lord Shiva.

It was a pleasant homestay in Gaichwan. The next morning, we spent 2 h at the Karna Temple, which delayed our departure for the next leg of the journey. After Netwad, the road conditions worsened. Our last stop was in Osla village, the first village in the Tons River basin. Between Netwad and Osla, there were six stops: Kot village (1770 m), Sidri (1850 m), Sankari (1940 m), Taluka (2110 m), Dharkot (2530 m), and Gangad (2300 m). This road runs along the left bank of the Supin River, which is the main tributary of the Tons River. The Supin River originates from Bandarpunch Mountain Peak. The other tributary, the Rupin River, originates in the neighboring state of Himachal Pradesh, with most of its catchment area falling within the state.

It was a heavy rainfall with a prediction for continue rainfall in the coming days. Initially, no one was ready to drive from Netwad. After an hour wait, a driver, having Bullero Taxi, was ready to drop us up to the last station. Netwad to Sankari was 12 km, passing through Kot and Sidri villages, whereas reaching to Taluka, it was another 12 km. There was an inundated stream on the way and the bridge on it was already washed away due to heavy downpour last night. The streams were overflowing. The taxi had to cross the stream, and the water level was rising. With great difficulty, the driver managed to cross it. After reaching Taluka, the driver refused to drive further, because, the road condition was worsened. We wait for about 2 h in Taluka to move further. A forest guest house is located in Taluka settlement in a serene landscape of the Supin River. We went to the guest house. In the meantime, forest officials alerted us not to go Hari Ki Dun, because we did not have permission from the State Forest Department. In Taluka, only a few tea stalls and Dhabas were available, no settlements and no agricultural fields. It is surrounded by dense deodar trees. From Taluka, the road goes upwards and it was still under construction. In many perennial streams, bridges were washed away. So, to cross the violent streams, temporary arrangements were made.

From Netwad to Sankari, apple orchards lined both sides of the road. The road runs along the course of the Supin River, creating a panoramic landscape. Numerous waterfalls, gorges, river meanders, and river terraces contributed to the spectacular landforms. Walnut trees could be seen along the roadsides, while the other side of the road featured steep mountains, with only grasses growing. A few trees from temperate forests dotted the steep slopes (Sati 2023b). The journey between the two villages was enjoyable. Both Kot and Sidri villages are known for their temperate climate and high apple production, with the fruits being of excellent quality. The settlements were beautifully constructed, with spacious, wooden, and brightly colored houses enhancing the beauty of the area.

Sankari is a beautiful village and a service center, serving as both a cultural and educational hub for the region. The village is home to the beautiful Someshwar Mahadev temple, which attracts many visitors from the Tons basin every year. In 2011, Sankari had a population of 270 people across 77 households. Although it is a small settlement, it boasts several quality homestays and numerous restaurants that serve delicious traditional dishes. As a result, tourists and trekkers often choose to stay in Sankari. The village is the starting point for the trek to Kedar Kantha. Since the Govind Wildlife Sanctuary is located in the Tons River basin, Sankari is surrounded by lush green temperate coniferous forests, teeming with rich biodiversity, both faunal and floral. The Tons River flows just below the village, adding to the beauty of the landscape. During the winter season, Sankari becomes very cold, as the surrounding areas receive heavy snowfall, and occasionally, the village itself experiences snow.

Sankari is a junction leading to Hari Ki Dun, Kedar Kantha, and Jakhol, the first village of the Rupin River basin. Jakhol is one of the largest villages in the Tons River basin, with a total population of 1601 people and 295 families. The village is located about 19 km from Netwar along the Rupin River. Its landscape is spectacular, and it experiences snowfall during the winter. At the center of the village lies the famous Someshwar Mahadev Temple. The houses are beautifully constructed using deodar wood and stones. A trekking route from Sankari passes through Jakhol village and leads to Devkyara. Jakhol village boasts a rich culture and cultural heritage, with centuries-old practices in agriculture and livestock farming. Rajma (kidney beans) and potatoes are the main crops grown here (Fig. 3.9). The villagers rear cows, goats, sheep, lambs, and horses. During the summer, animal herders migrate to alpine pasturelands for grazing.

Finally, we managed to find a large vehicle at Taluka. Many passengers were already aboard, but we managed to share seats and began an extremely challenging journey. The condition of the road was dreadful. We moved forward, swaying and jolting, with our hearts in our mouths. At one point, the vehicle almost overturned. The driver eventually stopped at Dharkot due to the terrible road conditions. The road was muddy, slippery, and under construction from Taluka to Gangad, with several bridges over the violent streams yet to be built. Vehicles often traveled only as far as Dharkot, as the route to Gangad was considered highly dangerous. A high-volume, high-velocity stream joins the Supin River from its left bank, and vehicles

Fig. 3.9 Two crops—Rajma and potato are growing in a cropland near Gangad village. Photo By author

must cross it to reach Gangad. During heavy rains, it is almost impossible to cross the stream without a bridge. From Dharkot to Gangad was approximately 3 km, and we had no choice but to walk. The path was steep and narrow, and to make matters worse, it was raining heavily. It felt as though the mountain was collapsing on us. Somehow, we managed to reach Gangad village (Fig. 3.10).

Gangad (2300 m) is a picturesque settlement located on the right bank of the Supin River. It is connected to Dharkot by a bridge. The village has a total population of 535 people, and the houses are densely clustered. These homes are beautifully constructed using deodar wood and stones. Small patches of terraced fields are situated on both sides of the Supin River, where traditional crops such as millets (Manduwa), rajma, potatoes, and fafara are cultivated. Some families rear cows, while goats, sheep, and lambs are sent to Har Ki Dun for summer grazing.

As soon as we reached Gangad, it began raining heavily. Feeling hungry, we ate some Maggi before continuing our journey upward, as our final destination was Osla village. Along the way, we joined a group of women from Osla village and proceeded along the banks of the Supin River. The path was rocky, uneven, and involved a steep climb. In fact, there were no proper trails—only deep ditches and large stones.

The Supin River was in its youthful stage, roaring fiercely. Some travelers informed us that there had been heavy rainfall in the river's catchment area for the past few days (Fig. 3.11). After walking a little further, we crossed the river via a hanging bridge that was partially broken, risking our lives (Fig. 3.12). We hurried across the bridge, and as soon as we reached the other side, a wave of fear swept over us. We wondered whether the bridge would still be there when we returned, given its terrible condition. For the locals, however, this was part of their daily routine. The real dilemma was that, even after so many years of India's independence, a proper bridge had not been constructed, and the residents continued to suffer. The real climb began after this. The ascent was steep, rugged, and stony, making it extremely difficult to traverse. As soon as we placed 1 foot forward, the other

Fig. 3.10 (**a**) Gangad villages is located on the right bank of the Tons River making a beautiful rural landscape, (**b**) a wooden-made beautiful building is used for homestay tourism in Gangad village. Photo By author

Fig. 3.11 The Supin River is flowing in its youthful stage, very violent, near Osla village. Photo By author

Fig. 3.12 A wooden bridge nearly broken on the way to Hari Ki Dun. Photo By author

seemed to slide back automatically. After much effort, we finally reached the outskirts of the village. The intensity of the rain was gradually increasing, and somehow, we managed to find shelter in a house for an overnight stay. As soon as I entered the house, I saw an elderly woman who struck me with a wave of nostalgia. Old memories came flooding back as she reminded me of my grandmother. The woman looked at me hesitantly, and I couldn't help but notice the uneven streets of the village, the house's terrace, and a few cows tied at one end of the house—everything was reminiscent of my childhood. But I quickly gathered myself. It was just a fleeting dream. I walked onto the house's terrace, placed my bag on a bench, and sat down to catch my breath.

Someshwar Mahadev is the presiding deity of the village. At the center of the settlement stands the temple of Lord Someshwar, which opens only on the 20th day of the Bhadra month. Despite being surrounded by water, the village suffers from a severe shortage of drinking water. A natural spring exists in the middle of the village, but it is not capable of providing potable water and is surrounded by heaps of filth. The village streets are similarly unclean and cluttered. During the rainy season, many villagers move to Har Ki Dun with their sheep and goats for summer pasturing, which leads to a severe shortage of milk in the village. As a result, the locals rely on Amul milk. The villagers are strong and accustomed to the rugged terrain. They often trek 14 km to Har Ki Dun to provide salt to their animals, completing the journey within half a day. The village has only one primary school, while the nearest high school is located some distance away. Consequently, most of the current generation has only a high school education. However, in recent years, with support from the state government, some children have started traveling to Dehradun for higher education. Electricity is a rare commodity in the village. Most of the time, the village is shrouded in pitch darkness, with only one solar-powered bulb lighting up a small area. Nights are cold due to the high altitude. Dinner consisted of ghee, curd, roti, Rajma, rice, buckwheat, amaranth, pumpkin leaves, and other leafy vegetables, all of which were delicious. After chatting for a while, we went to sleep around 11:00 pm. The village lacks essential facilities. The streets are narrow, dirty, stony, and uneven. There are no toilets, bathrooms, or other basic amenities. The village has no bank, post office, or communication facilities, and there is no phone service. Despite producing a significant quantity of potatoes and Rajma, there were no potatoes available for consumption. The situation in the village was both puzzling and distressing. It felt as though the residents were living in the seventh century, with little connection to modern advancements.

Osla village is located at an altitude of 2750 m, situated on terrain ranging from gentle to steep slopes. It marks the first settlement in the Tons River basin, beyond which the Greater Himalayas begin, leading into uninhabited no-man's land. The houses in Osla are constructed using deodar wood and stones, reflecting traditional Himalayan architecture. Cowsheds are integrated into the houses, showcasing the functional design suited to the region's climate and lifestyle. A few villagers are engaged in weaving traditional clothes. They extract wool from goats, spin it into thread, and weave garments using traditional handlooms. Rajma and potatoes are the primary crops cultivated in the village, along with green vegetables and millets. Rajma, in particular, serves as a commercial crop due to its non-perishable nature. It can be stored within households for extended periods, allowing farmers to sell it when market prices are favorable and road conditions permit transportation.

However, selling potatoes is challenging due to inadequate market facilities and poor connectivity. To reach Dharkot, villagers must walk approximately 7 km each way, navigating poorly maintained and rugged trails. The condition of these trails poses significant difficulties for both travel and transportation.

The next day, early in the morning, we set out for the Har Ki Dun trek (Fig. 3.12). It had been drizzling since morning, with occasional heavy rain. Despite the weather, we gathered our courage and moved forward from Osla village. The path was terrifying, with small streams overflowing, making it difficult to take even a single step. Still, we kept moving forward. The closer we got to the Himalaya, the more beautiful nature became, and divine waves seemed to spread throughout the region. The literal meaning of Har Ki Dun is the Valley of Vishnu. We felt as if we were getting closer to God. After covering a distance of about 4 km, we reached a campsite, a small Bugyal, below which the Supin River flows (Fig. 3.13). In the Himalaya, alpine grasslands are called Bugyals. This particular Bugyal, located a little distance from the Supin River, was very attractive. Our hearts rejoiced, and our minds were eager to merge into the lap of nature. We moved ahead, even though the path to the water source was difficult. As time passed, the weather worsened. We had covered about 16 km of walking. We continue proceeding toward Hari Ki Dun with steep slopes. The trail was up and down. On the way, there were numerous waterfalls, rivulets, streams, cascades, and gorges. The trail was becoming difficult as we were moving upwards. We had to find our own shelter. Our bodies were exhausted, and there was no shelter for rest anywhere.

Fig. 3.13 The Supin River is flowing through a deep valley near Osla village making spectacular river valley landscape. Photo is taken from the Campsite (3000 m), a Bugyal, 4 km from Osla village on the way to Hari Ki Dun. Photo By author

Hari Ki Dun, nestled in the lap of the Himalaya at an average altitude of 3500 m, is part of the Govind National Park and Wildlife Sanctuary (GNP&WLS). The valley is long and expansive, surrounded by alpine grasslands, with the Supin River—a major tributary of the Tons River—flowing through its center. The GNP&WLS covers an area of about 958 km^2, including the upper Tons River basin as well as the Rupin and Supin Rivers. The terrain is tough and rugged, making it one of the most remote regions in the Uttarakhand Himalayas. During the monsoon season, the area receives an average rainfall of 1500 mm, which poses significant challenges for trekking. Hari Ki Dun is rich in biodiversity, home to more than 50 floral species, 70 species of butterflies, 20 species of mammals, and 100 species of birds (Bhardwaj et al. 2012; Sathyakumar 1994). The valley is also adorned with a variety of flowers and medicinal plants, adding to its natural allure. The State Forest Department maintains a guesthouse in the valley, but it is old and poorly maintained, with no caretaker present. During the monsoon season, staying overnight is particularly difficult, and visitors are generally not permitted due to adverse weather conditions. When we arrived at the guesthouse, the sun was setting, and the valley was quickly shrouded in darkness as thick clouds covered the entire area. Visibility dropped to zero, and we found ourselves without food, light, or any means of communication. The cold weather added to the challenges of the night. After chatting for a while to pass the time, we retreated into the guesthouse, not realizing when we eventually fell asleep. The next morning, taking a bath was out of the question as the weather had worsened further. Shortly after sunrise, it began to rain again. However, the valley looked stunningly fresh, with the sun's rays illuminating the snow-capped mountain peaks and casting shadows in the Supin Valley, creating a breathtakingly panoramic view. Although we longed to stay in the serene beauty of Hari Ki Dun

Fig. 3.14 Hari Ki Dun Bugyal, the Supin River is flowing in between two mountain slopes. Photo By author

(Fig. 3.14), the journey ahead awaited us. Reluctantly, we began our return trip, carrying with us the unforgettable memories of this remarkable valley.

The whole day, we continued trekking back. It was evening, and we were considering staying overnight in Gangad. With some time left, we thought of traveling a bit further. There were two routes to reach Dharkot village: a steep climb and another along the roadside. The roadside route was longer and had landslides in several places. After a steep climb, we continued on the road, which had a slight slope, making it easier to walk. Fearing it might rain, we moved quickly. After about an hour, we reached Dharkot village (Fig. 3.15), where we found a tea shop on the road and decided to rest. Two of my companions were looking for a house to stay overnight. As time passed, just as we were about to follow them, it started pouring heavily. We ran to take shelter in a house where our two companions were already staying.

The house was made entirely of deodar wood, beautiful but with very low eaves. I bumped my head three times. In the entire Supin Valley, houses were made of wood and had unique designs. The hospitality of the villagers won our hearts. We sat on the "Tyabari" for a while and moved inside when the rain intensified. The dinner was delicious, and I tasted Fafda vegetable curry for the first time, along with buttermilk and Manduwa roti. Everything was wonderful. It was 11 pm, time to sleep, but the real challenge was the bathroom. It was the dead of night with heavy rain, and the bathroom was dirty and far away. It felt like the sky was falling. I realized that the Har Ki Dun trek was easier than dealing with Dharkot's bathroom. The night passed, and morning came. After breakfast, we had to continue our journey.

Fig. 3.15 Dharkot village (2530 m), located on the trekking route to Hari Ki Dun, where several Homestay are being constructed. Photo By author

The villagers were simple and kind. Initially, they said a vehicle was going, but then they refused.

Dharkot (2530 m) is a beautiful village with terraced agricultural fields on gentle slopes. The village is connected by road, but the condition of the road is critical, and it remains closed during the monsoon and winter seasons due to heavy rainfall and snowfall, respectively. Dharkot has a temperate climate—summers are pleasant, while winters are snow-capped. Additionally, due to its proximity to the Greater Himalayas, cold waves sweep across the region during winter. The village has fertile land, with organic manure used for cultivation. The main crops grown include wheat (after summer), red rice, potatoes, fafara (Fig. 3.16), Manduwa, rajma, apples, apricots, pears, peaches, and walnuts. Rajma and potatoes are sold in local and regional markets. The village is also developing homestay tourism, with tourists trekking to Hari Ki Doon often staying overnight in the village. Farmers rear animals such as cows, goats, sheep, lambs, and horses. Cows remain in the village and are stall-fed, while goats, sheep, lambs, and horses migrate to the higher slopes (alpine pasturelands) for summer grazing. Traditional foods, including milk and milk-based products like curd, ghee, butter, and buttermilk, are served to homestay tourists. The forest landscape is spectacular, with temperate forests dominated by deodar trees. Above the forest line, there are alpine grasslands, and beyond that, the snow-capped Greater Himalayas. It was observed that due to the high availability of deodar wood, most houses are constructed using deodar wood. People worship Someshwar Dev on various occasions. Every male member has two or more wives.

Fig. 3.16 Fafara, a green vegetable and traditional millet, is grown largely along the Hari Ki Dun Trekking mainly in the high altitude. It is highly nutritious crop. The photo presents a Fafara field in Dharkot village. Photo By author

The owner of the homestay has two wives, and both wives have three children each. They live together happily.

We started trekking along the road. There were landslides everywhere, and streams were overflowing. After walking about 8 km, we saw a tractor. We talked to the driver, and he agreed to take us a short distance, about 4 km. Ahead, a stream was overflowing, and the bridge was broken. The villagers had placed a thick log over the stream for crossing it. We gathered our courage and crossed the stream using the log. Then we moved ahead. The road to Sankari was terrible. The worst part was when the road was completely broken, with big rocks falling from the slope above and a huge rainy stream ahead. It was like being between a rock and a hard place. Our companions were scared. A big tree had fallen on the road. After some time, the tree was removed. We had to cross a stormy stream ahead. Some laborers came forward to help us. We took off our shoes and threw them to the other side. I still had my backpack on. We barely managed to cross the stream, risking our lives. Then we moved ahead. We changed two vehicles before reaching Sankari.

Now the road was easier. We reached Mori via Netwar and decided to visit Hanol, and abode of Mahasu Dev, the god of justice. The village of Hanol, situated on the banks of the Tons River, is very ancient and famous for its cultural heritage. Hanol is located both sides of the Tons River, making a very panoramic view. The Tons River flows down making scarf and two parts of village are situated on the two river terraces. The road between Mori and Hanol traverses along the left course of the Tons River on gentle slopes, both sides of which are dense pine trees. The area is lush green with several patches of fertile arable land where paddy, Mango, and seasonal vegetables are grown abundantly. A 30 km journey gives you peace and enjoyment. Since Hanol is located in the valley at an altitude of 1120 m (average), tourists and pilgrims visit the village throughout the year. We had a delicious meal in Hanol, then returned to Mori and reached Purola through the dense pine forests, where we stayed overnight. The next morning, we surveyed two villages and reached Dehradun in the afternoon. Thus ended a geographical, cultural, and adventurous journey. This trip will always remain in my heart.

Here, we further describe the land, people, and culture, including folk deities and folklore, songs and dances, as well as the economy and development. Additionally, we provide suggestions for developing the Tons River basin in terms of its economy, culture, and society.

3.3 Land, People, and Culture

This region is known by three names: Jaunsar and Bawar, Rawain, and a small part of Western Garhwal. Jaunsar and Bawar fall within the Chakrata Taluk of the Dehradun district, in the Tons River Basin. It starts from Kalsi, a service center and village, near the mountains and close to the Tons River, which joins the Yamuna River. Most of the area consists of barren grassland up to Chakrata, with small patches of arable land and forests. Beyond Chakrata, dense temperate forests extend to Tyuni. At middle altitudes, pine forests are common, and at higher elevations, extensive alpine pasturelands are found. Above these pastures are the snow-capped peaks of the Himalayas. Arable land is primarily found along the Tons River, streams, and water bodies. The second region, Rawain, has more fertile arable land. It begins at the Yamuna Bridge, about 30 km from Mussoorie. Initially, the landscape is dry and barren, with sparse pine forests and scrubland. As the area ascends toward Yamunotri and then turns toward Purola before Bhatwadi, the landscape becomes beautiful, with dense forests and large arable areas. This region is called Rawain. A road runs along the Kamal River, a tributary of the Yamuna River. The terrain is gentle, and the Kamal River creates a vast fertile valley. The upper edge of this valley is known as Rama Sarain, famous for its high production and productivity of red rice. The road continues through dense pine forests to Mori, situated at an altitude of 2500 m. The path ascends initially and then descends. Mori is an urban settlement on the banks of the Tons River and is known as the gateway to Hari Ki Dun. From Mori, roads branch off in four directions: Hari Ki Dun, Hanol, Tyuni, and Purola, making it a junction in the Rawain region. The third region is Western Garhwal, also called Shiktur Patti, and is the most beautiful of the three. It is characterized by dense temperate forests, primarily deodar, alpine grasslands, waterfalls, rapids, and streams. The high Himalayas, which remain snow-capped throughout the year, form the region's last section. In the alpine pasturelands of Hari Ki Dun, a variety of flowers bloom during autumn when the snow melts, and medicinal plants also thrive during this season. The region remains snow-capped in winter.

The Tons River and its major tributaries—Rupin, Supin, and Pabbar—form a comprehensive drainage system and create a beautiful river landscape. Bandarpunch is the highest mountain peak, beyond which lies the Tibet Autonomous Region of China. To the west of this peak is Himachal Pradesh. The Tons River is formed at Netwar, where the Rupin and Supin rivers meet. These glacier-fed rivers, in their youthful stage, create stunning river landscapes. The region is characterized by deep river valleys, gorges, rivulets, waterfalls, cascades, and river terraces, all shaped by the Rupin, Supin, and Pabbar rivers. During the rainy season, these rivers become highly volatile, leading to flash floods, debris flows, and landslides. Every year, the region experiences severe catastrophes during the monsoon season.

The Jaunsaris inhabit the Chakrata Tehsil of the Dehradun district. This region was historically known as Panchaldesh, a name derived from the Pandavas. In the Tons valley, the Jaunsaris consider themselves descendants of the Pandavas, while the Bawaris trace their roots to the Kauravas. The Jaunsaris speak Jaunsari, a language distinct from both Garhwali and Kumaoni. Throughout ancient times, this region was ruled by various dynasties, including the Sirmaur dynasty, Jubbal, Ashoka, and the Pandavas.

The culture, customs, and rituals performed by people are quite different. Polygamy and polyandry systems were practiced in the past. The rich tribesmen tended toward polygamy, whereas the others engaged in polyandry system, which is also known as Fraternal Polyandry where two or more brothers share a wife. While in Jaunsar, a woman has two or more than two husband, a man has two or more than two wives in Rawain region. However, these practices are gradually declining, giving way to monogamy (Berreman and Gerald 1963; Majumdar, 1962). On the way to the Hari Ki Dun, we stayed in a homestay in Gaichwan village. The homestay was owned by Mr. Rangad, who was a retired government official. While discussing with him about the region, he seemed not happy because he could not marry with another woman. He was referring that his friends have two wives and he is unlucky. On the other hand, while we were trekking to Osla village, 5 km from Gangad village, three women were accompanied us and all were the second wives of their husband. However, they were happy, living in the joint family together.

Remarriage and widow marriage are not considered taboo, although the groom is required to pay a dowry, known as the "marriage tax." In the case of remarriage or widow marriage, the groom must pay double the usual amount. The Jaunsaris follow the practice of bride price, believing that parents invest significantly in raising, educating, and caring for their daughters. When the daughters are capable of earning money and supporting family livelihoods, they get married. Daughters are considered valuable assets as they help care for the family and carry out farm work. As such, a bride price is levied when a man wishes to marry a woman. Divorce is also not taboo and is generally accepted in society, as is remarriage. If a divorced woman returns to her paternal home, her family must reimburse the bride price to the man's family. Additionally, if a woman divorces her husband to marry another man, the second husband is required to pay double the bride price to the first man's family. The Rawain region extends across the Tons River and Kamal River valleys in the Uttarkashi district, as well as the Jaunpur area in the Tehri district. On the opposite side of the Tons River is Himachal Pradesh, which is similarly influenced by Rawain culture. The caste system in the region consists of three categories: landowners, service-oriented individuals (including artisans), and Brahmins (Hindu priests), and these categories are determined by birth.

The people of the region are striking, tall, and have a fair complexion, with women being especially attractive. In the past, it was believed that women could hypnotize and control a handsome person by serving food mixed with butter. The culture, customs, and rituals of the region are unique and distinct from other parts of the Himalayan region. Their language, food habits, and clothing differ from those of surrounding areas. The language spoken is primarily Pahari, influenced by Himachali, and is known as Jaunsar and Rawain. It varies according to the different drainage basins. Millets are a staple food, and people commonly eat Manduwe ki Roti with a traditional curry made from seasonal green leaves, such as Fafara, Chaulai, and Pumpkin. Cooked red rice and Rajma curry are also popular dishes. Milk and dairy products, such as butter, buttermilk, ghee, and milk, are key ingredients in their main meals. Women wear a Ghaghara with a white shirt and a Chunni during the summer, and in winter, they wear a long woolen coat to protect themselves from the cold wind and cool climate. Men wear a Kurta and Pajama, along with a long stole around their neck, complemented by woolen clothes. Thalka or Lohiya is a long woolen coat worn by both women and men, primarily during the winter when cold waves sweep through the region. On auspicious occasions, people wear gold and silver ornaments. Popular gold ornaments include Nath, Kaan Phool (earpieces), and Mala, Munra, and Soota (necklaces). Other gold ornaments include

Jhappi, Globand, Dhandrahar, Paunchi, Atardan, Shiv Sangal, and Sootu. Widows generally do not wear these ornaments. Men wear long Kadas (bracelets), large rings, and earrings.

3.4 Folk Deities

Mahasu Dev, Someshwar Mahadev, Pokhu Devta, Karna, and Duryodhana temples are located in the Upper Tons River region. These are folk deities worshiped in the area. The four temples depicted in Fig. 3.17 show that the architecture and materials used in all of them are quite similar. These temples are primarily constructed from wood, and the architectural style is consistent across them. The temples of Pokhu Devta and Karna are well-maintained, whereas the other two temples, Duryodhana and Someshwar, appear old and in a state of disrepair.

People worship both the Kauravas and Pandavas, including Karna. A temple dedicated to Karna is located in Dewera village near Netwad. Karna is worshiped throughout the year, especially during times of drought, when people pray to him for rain. Karna is believed to be the God of Rain. During the Dvapara Yuga, this

(a) (b) (c) (d)

Fig. 3.17 (**a**) Duryodhana temple in Mori, (**b**) Pokhu Devta temple in Netwad, (**c**) Karna temple in Dewera, and (**d**) Someshwar temple in Osla. Photo By author

region was ruled by both the Kauravas and Pandavas, and the Pandavas are believed to have lived here during their Vanvas (exile) and Agyatvas (period of incognito). As a result, the people of the region feel a strong connection to them. It is common for people living in the highland areas to worship the Kauravas, particularly Duryodhana, while those in the river valleys worship the Pandavas. The practice of polyandry and polygyny in this region is believed to be influenced by the customs of the Pandavas.

3.5 Shiva and Shakti

Shiva and Shakti are the supreme folk deities in the Himalayan region. The Himalayas themselves are believed to be the embodiment of Lord Shiva. Prominent pilgrimage sites like Kailash-Mansarover, Chhota Kailash, Kedarnath, and Jageshwar are considered sacred abodes of Lord Shiva and Parvati. In the Uttarakhand Himalaya, temples dedicated to Shiva and Shakti are found throughout the region, across all river basins. In the Tons River basin, temples are devoted to various folk deities such as Pokhu Devta, Someshwar Devta, Karna, and Duryodhana. However, the presence of Lord Shiva and Shakti, along with their idols and Lingams, is always evident. These folk deities are believed to be manifestations of Shiva and Shakti.

3.6 Duryodhana

Duryodhana is worshiped by 22 villages of Fateh Mountain shrine, Panch Gai, Osla-Netwad, and Hari Ki Doon. It has many temples located along the Supin River valley. The popular temples are located in Mori, Netwad, Saur, Jakhol, Datmir, Gangad, and Osla. It is believed that Duryodhana protected the people of Tons River valley in many occasions and therefore, folks worship him as their deity. A festival is celebrated in Vaisakh to worship Duryodhana and Kauravas.

Duryodhana, a prominent figure in the Indian epic Mahabharata, is uniquely revered in the Tons River valley. This practice stands out as unusual in Indian cultural and religious contexts, where Duryodhana is often portrayed as the antagonist. The Mahabharata has numerous regional adaptations, and in some versions prevalent in the Tons River valley, Duryodhana is depicted in a more favorable light. He is sometimes seen as a brave and just ruler rather than a villain. Local legends suggest that the inhabitants of certain villages in the Tons River valley consider themselves descendants of the Kauravas, Duryodhana's family. This lineage is a source of pride and cultural identity. Duryodhana's valour, loyalty to his friends, and determination are qualities that resonate with the local communities. His unwavering loyalty to his friend Karna is particularly admired. The Tons River valley, being a remote and rugged region, has a culture that values strength, resilience, and loyalty, traits that are embodied in the character of Duryodhana. Several temples in the region are dedicated to Duryodhana, where he is worshipped as a deity. These temples are centers of local religious life and community activities. Annual festivals and rituals are held in honor of Duryodhana, attracting devotees from surrounding areas. These celebrations often include traditional music, dance, and theatrical performances depicting episodes from the Mahabharata. The oral traditions and folklore of the Tons River valley often highlight Duryodhana‘s positive attributes. These stories have been passed down through generations, shaping the collective memory and cultural practices of the region. Worshipping Duryodhana serves as a unifying factor for the communities in the Tons River valley. It reinforces a shared identity and cultural heritage. The temples and festivals dedicated to Duryodhana also boost local tourism and economic activities, providing livelihoods for many residents. The remote location of the Tons River valley has allowed unique cultural practices to develop and persist, relatively untouched by mainstream religious narratives. This isolation has helped preserve the distinct worship of Duryodhana.

3.7 Pokhu Devta

The Pokhu Devta temple is located on the left bank of the Tons River in Netwad village, Uttarkashi district. Pokhu Devta is a popular folk deity, known as the god of justice. It is believed that anyone who visits the Pokhu Devta temple and worships sincerely will have their problems resolved. In the temple premises, devotees are required to strictly follow the rules and regulations. The people of the region have deep faith in this folk deity, who is regarded as the king of the area. Local folklore holds that the head of Pokhu Devta resides in Pataal Lok (the underworld), while the upper part of the body is on the earth. According to legend, in ancient times, a monster named Kirimar used to terrorize the local people. Duryodhana fought and defeated the monster to protect the people. After cutting off its head, Duryodhana threw it into the Tons River. Remarkably, the head began to flow upstream, eventually coming to rest at the confluence of the Rupin and Supin rivers. It was at this spot that Duryodhana is said to have built a temple, which is now known as Pokhu Devta‘s temple or Pokhvir.

The Pokhu Devta temple is unique in many ways. Devotees and even the priest are unable to see the idol of the deity. Despite this, the faith in Pokhu Devta remains unwavering. Interestingly, the deity is worshipped through symbols like a knife and sword. The idol of Pokhu Devta is positioned in an unusual manner—its head is down and its legs are up, with the idol being naked. As a result, devotees do not directly view the idol. It is believed that Pokhu Devta provides protection and assistance to the people of the region during times of adversity.

A fair is organized every November at the temple. During the fair, the village priest predicts the future of the village, including matters like prosperity and crop yields. These predictions are generally believed to come true, adding to the temple's mystique. Because of these unique characteristics and beliefs, Pokhu Devta‘s temple has become a renowned pilgrimage site, attracting tourists from far and wide. In the first chamber of the temple, there are dried bloodstains on the sacrificial altar. Inside the temple, a Shiva Lingam is installed, with the chamber of Pokhu Devta positioned behind it. The face of Pokhu Devta remains hidden from view, creating a fearsome aura. Consequently, both the priests and devotees worship with their backs turned to the deity. Pokhu Devta is considered to be a servant of Lord Shiva. According to ancient scriptures, Pokhu Maharaj is regarded as a representative of Karna and a servant of Shiva, known for his fearsome appearance and strict demeanor toward his followers. His domain is free of thefts and crimes, further emphasizing his protective nature.

3.8 Karna

The Karna Temple, nestled in the serene Tons River valley, is a revered shrine dedicated to Karna, the legendary warrior of the Mahabharata. This sacred temple not only honors Karna‘s extraordinary bravery, generosity, and unwavering loyalty but also stands as a symbol of the region's rich cultural and spiritual heritage. Located within the scenic Tons River basin, the temple draws both devotees and tourists, captivated by its deep spiritual significance and historical importance. Karna, a central figure in the Mahabharata, is celebrated for his valour and his steadfast adherence to principles, even in the face of adversity. The temple's presence in this picturesque valley highlights the close connection the local communities share with the epic and its key figures. Architecturally, the Karna Temple reflects the traditional Himalayan style, with its wooden framework, intricate carvings, and sloping roofs designed to endure the heavy snowfalls typical of the region. The sanctum of the temple houses an idol of Karna, adorned with offerings from the devotees, including traditional ornaments and floral garlands.

A highlight of the temple's cultural significance is the annual Karna Jayanti, a festival celebrating the birth anniversary of Karna. During this time, devotees gather for special prayers, rituals, and cultural programs, celebrating Karna's virtues and his embodiment of selflessness and duty. The temple also serves as a space for community gatherings, spiritual discourse, and cultural events, fostering unity and a sense of continuity within the local population.

The Karna Temple plays a pivotal role in promoting tourism in the Tons River valley. Its spiritual significance draws not only pilgrims but also history enthusiasts, cultural explorers, and nature lovers. The temple's strategic location makes it a key point for various trekking routes and outdoor activities, offering visitors a gateway to explore the valley's breathtaking natural landscapes and rich cultural heritage. As a hub for tourists, the temple connects travelers to the surrounding beauty, with its scenic surroundings and nearby attractions such as alpine meadows, dense forests, and majestic mountain views. Whether visiting for spiritual reasons, historical interest, or the allure of outdoor adventures, the Karna Temple serves as a central attraction in the Tons River valley, helping to promote sustainable tourism and local economic development.

3.9 Someshwar Dev

Another folk deity is Someshwar Dev, believed to be the incarnation of Lord Shiva. Someshwar Dev has temples in many places in both the Rupin and Supin River valleys. The famous ones are located in Jakhol and Osla villages. A cultural procession of Someshwar Mahadev takes place occasionally, when the idol of Someshwar Mahadev is carried from village to village (Sati 2023c).

Someshwar Devta, often associated with Shiva, the Hindu god of destruction and transformation, is believed to be the protector of the valley and its inhabitants. The devotion to Someshwar Devta is evident through the numerous temples dedicated to him, scattered throughout the region. These temples serve as focal points for community gatherings and religious ceremonies, particularly during festivals and special occasions, when the deity is celebrated with elaborate rituals, music, and dance. One of the key aspects of worshipping Someshwar Devta is the traditional Jagar ceremonies, which involve invoking the deity through folk songs and dances performed by local musicians and storytellers. These ceremonies are integral to the cultural fabric of the Tons River valley, preserving ancient traditions and fostering a sense of community among the residents. The influence of Someshwar Devta extends beyond spiritual protection, as he is also considered a guardian of the natural resources and agricultural practices in the region. Farmers and herders often seek the blessings of Someshwar Devta for good harvests and the well-being of their livestock. This connection between the deity and the agricultural livelihood of the people underscores the harmonious relationship between culture, religion, and nature in the Tons River valley (Sati 2023d). In recent times, the reverence for Someshwar Devta continues to thrive, blending traditional beliefs with contemporary practices.

3.10 Mahasu Dev

Mahasu Dev, believed to be an incarnation of Lord Shiva, is also known as the God of Justice. Most of the disputes among the villagers are resolved at the village level by the head of the village in the name of Mahasu Dev. The temple of Mahasu Dev is situated on the left bank of the Tons River, on a river terrace surrounded by rural settlements and business avenues (Fig. 3.18). The village is named Hanol. It is beautifully constructed using traditional architecture. Similarly, there are many other temples dedicated to Mahasu Dev located in various places within the Jaunsar-Bawar cultural realm.

Mahasu Devta is a highly revered deity in the Jaunsar-Bawar region of the Garhwal Himalaya, particularly in the Tons River valley. Known as the principal deity of the local communities, Mahasu Devta's worship is deeply ingrained in the cultural and spiritual fabric of the region. According to local legends, Mahasu Devta is one of the four brothers, divine incarnations of Lord Shiva. The four

Fig. 3.18 A very spectacular view of the divine site, the temple of Mahasu Devta in Hanol village. Photo By author

brothers—Basu, Pavasi, Chalda, and Botha—collectively known as Mahasu Devta, were sent to the region to protect the people from evil forces and ensure justice and prosperity. Mahasu Devta is considered the protector and preserver of the local populace, invoked for blessings, protection from natural calamities, and guidance in personal and communal matters. The reverence for Mahasu Devta underscores the strong bond between the deity and the daily lives of the people. The main temple dedicated to Mahasu Devta is located in Hanol village, nestled in the Tons River valley. This temple is an architectural marvel, showcasing traditional wooden construction with intricate carvings and decorations. Several other temples dedicated to Mahasu Devta and his brothers are scattered across the region, serving as important spiritual centers. One of the most significant festivals is the Mahasu Jatar/Jatar, held annually in August–September. This grand event attracts thousands of devotees who participate in rituals, processions, and cultural performances. The fair includes traditional music, dance, and a vibrant display of local customs. The deity's oracles, known as "pujari" or "malis," play a crucial role in conveying divine messages and guidance to the people. These oracles, believed to be possessed by the spirit of Mahasu Devta during rituals, provide counsel on various matters affecting the community. Mahasu Devta's blessings are sought for agricultural prosperity. Rituals and offerings are made before the sowing and harvesting seasons to ensure good crops and protect against natural disasters. The deity's influence extends to social customs and legal matters. Disputes within the community are often resolved in the presence of Mahasu Devta, with the belief that divine justice will prevail. This practice reflects the deep-seated trust in the deity's fairness and wisdom. The worship of

Mahasu Devta is closely tied to the natural environment of the Tons River valley. The deity is considered a guardian of the forests, rivers, and wildlife, emphasizing the harmonious coexistence between humans and nature. The spiritual practices associated with Mahasu Devta include meditative and ascetic traditions that reflect the profound spiritual heritage of the region. Devotees engage in various forms of penance and spiritual discipline to seek the deity's blessings.

3.11 Major Fairs and Festivals

The cultural heritage of the Tons River basin is rich and vibrant, with people performing various fairs, festivals, customs, and rituals that are unique to the region. Some of these fairs and festivals are exclusive to the area, as described below.

3.12 Bissu Mela

The Bissu fair is a significant cultural identity of the Jaunsar-Bawar region, symbolizing the inseparable bond between humans and their surroundings. It serves as a platform to share stories, traditions, and celebrate the community's agricultural practices. Celebrated with the onset of the spring season, the fair is dependent on agriculture and farming. The exact origin of the fair is unknown, but it marks the beginning of the good harvest season, typically celebrated in the month of Chaitra during the Shukla Paksha. The Jaunsari tribes come together to worship Santura Devi, believed to be an incarnation of Goddess Durga. During the fair, people wear new traditional attire, sing folk songs, and perform folk dances, strengthening their cultural and social bonds.

3.13 Maroj or Magh Maroj

Maroj, one of the most important festivals in the Jaunsar-Bawar region, is celebrated during the Magh Mela. The festival is named after Maroj (ogre), a large, cruel, and frightening creature who terrorized the people of Jaunsar-Bawar for an extended period. To honor this folklore, people engage in the ritualistic sacrifice of Maroj to the folk deity during Magh Mela. The festival begins with the sacrifice of a goat at the Kaylu Maharaj temple in Hanol, and each family in the village offers a goat as a symbol of sacrifice to the folk deity. People residing in the river valleys of Jaunsar-Bawar and Rawain realms believe that they are descendants of the Pandavas, and they sacrifice the goat symbolically to represent the killing of Dushassan, who attempted to disrobe Queen Draupadi, the wife of the Pandavas. The meat (locally known as Banthi) is distributed among the village, with the remaining meat being dried traditionally for use during the winter, when the region experiences heavy snowfall. The month-long festival provides an opportunity for family members to reunite with relatives and friends, share meals, sing folk songs, and perform folk dances. The married daughter plays a central role in the celebrations, and Banthi is given to everyone after the gathering concludes. Folk dances like Haarul and Taandi are performed with the traditional beats of Dhole Damau during the fair. The festival coincides with a time when agricultural land is left fallow in preparation for sowing paddy, as people believe that keeping the land fallow enhances its fertility.

3.14 Budi Bagwal

Budi Bagwal is celebrated a month after the Deepawali festival in Uttarakhand. This 5-day celebration is unique for being an eco-friendly festival, as it is celebrated without the noise of firecrackers. Instead, firewood is burned for lighting, making it a quieter and more nature-conscious event. The legend behind Budi Bagwal relates to Bhima, the brother of the Pandavas. According to the tale, Bhima went into the forest and did not return during Deepawali. He only returned a month later, after which the Pandavas decided to celebrate the festival. During Budi Bagwal, people engage in the tradition of playing Bhaillyo, a ritual in which they burn wood, run across agricultural fields, and worship Mahasu Dev. Delicious foods are prepared and shared among the community, adding to the festive spirit of the occasion.

There are several other festivals celebrated in the Tons River valley, with the Maund festival, Nunai festival, and Jagra fair being among the most popular. The Maund festival, celebrated in May and June, is devoted to fishing, coinciding with the best time for fishing along rivers and streams in the Jaunsar-Bawar region. During this festival, people buy fishing equipment and engage in fishing activities, accompanied by traditional dances. The Nunai festival honors God Shilgur, the god of sheep, and is celebrated in Baila and Jadi in September. It is a festival deeply rooted in Jaunsari customs, reflecting the traditional culture of the region. The Jagra fair is held in Hanol at the Mahasu Devta Temple in the last week of August. It is a significant cultural event in the region, attracting many devotees and visitors to the temple premises for rituals, prayers, and celebrations.

3.15 Folk Dances

Performing dance and music is an integral part of life for the people of the Tons River valley, and they engage in these activities on all occasions. Fairs and festivals are vibrant, offering a rich reflection of the region's culture and traditions. Barada Nati, Harul, and Raso are popular folk dances in the region. Figure 3.19 depicts different forms of folk dance: Figure (a) shows people performing Bhaillyo, where they hold sticks in their hands while dancing during the Bissu fair. Figure (b)

(a) (b) (c) (d)

Fig. 3.19 People celebrating Budi Bagwal (**a**), performing sword dance, and performing folk dance during Bussi fair (**b**, **c**, **d**). Photo By author

portrays the Sword dance, performed by young men dressed in white clothes. Figure (d) showcases young women performing the Bussi dance. The dresses worn by the performers are colorful and traditional, with distinct attire for men and women.

3.16 Dev-Devi Naach

Dev-Devi Naach (Dance) is performed on various occasions, with the folk deities, primarily Lord Someshwar, Mahasu Devta, Karna, and Goddess Shakti, manifesting through human dancers. This process is called Dev Avataran and Dev-Devi Dance (Fig. 3.20). Traditionally, socially backward people play Dhole and Damau, the local musical instruments, while men and women perform the dance. The Dev-Devi dance is typically performed during many occasions, especially in the winter. Animal sacrifice, mainly goats, is practiced, as it is believed that the folk deities demand such offerings for the well-being of the people. Almost every temple premises hosts the sacrifice of goats to offer to the deities during the Dev-Devi Naach and while celebrating fairs and festivals. The Jaunsaris also believe in ghosts and demons, which are said to inhabit crossroads and rivers. They pray to these spirits for protection, and sometimes, these spirits are believed to manifest in human form and participate in the dance.

(a) (b)

Fig. 3.20 (**a**) A Jaunsari man playing Dhole. (**b**) People are performing Dev-Devi Naach, dancing around a fire pod, locally known as Dhuni. Photo By author

3.17 Economy and Development

The economy of the Tons River valley is primarily based on traditional subsistence agriculture, with millets and Red Rice being the most prominent crops. However, the production and productivity of these crops are limited, and with a growing population, the agricultural output is insufficient to meet the needs of the people, leading to food insecurity and malnutrition. Agricultural practices vary across different altitudes in the region. In lower altitudes, crops such as rice, tomato, and wheat are commonly grown, while apple, peach, plum, apricot, and walnut thrive in the middle altitudes. In the highlands, potato and Rajma are the main crops. Economically, residents of the lower altitudes tend to have higher incomes, and their livelihoods are more sustainable compared to those in the highlands. Livestock rearing and homestay tourism also contribute to the economy, though these activities are limited, with only a small portion of the population involved. Forest products, such as fuelwood, fodder, wild fruits, and medicinal plants, are abundant, but their utilization remains low. Pine resin is extracted in large quantities and sold to industries for use in making road tope, bangles, and various other products.

Jaunsar-Bawar and Rawain are among the most underdeveloped regions of the Uttarakhand Himalaya, particularly within the Tons River basin. Transportation is a major challenge, with many villages lacking any form of connectivity. For those that are connected by road, the condition is poor, and the walking trails are rough, requiring people to walk miles to reach their rural settlements from the road. A significant number of villages remain without electricity, and those with power supply face inconsistent service. Telecommunications are also limited, with the Netwad service center being the last point of connectivity, and only Jio works in the valley regions, though the network is unreliable. Agriculture is largely rainfed, as there is a lack of adequate irrigation facilities. The region is also devoid of industries, meaning that products are sold at low prices. It is evident that infrastructure is deteriorating, particularly as one moves from the river valleys to the highlands. Villages such as Dharkot, Gangad, and Osla are isolated and underdeveloped, with development seemingly bypassing these areas. The people in these remote locations appear to be living in conditions reminiscent of a past century.

3.18 Education

The region is educationally backward, especially in the highland villages. In Osla village, only a primary school is available, and the nearest high school is located 5 km downstream. Due to the undulating trail, children often avoid going to school. As a result, many people send their children to schools and colleges in Dehradun, Vikas Nagar, and Purola, with financial assistance from the state government. The entire region is designated as an ST (Scheduled Tribes) area, meaning that every household, regardless of caste, creed, or religion, is considered to belong to a Scheduled Tribe. This grants the tribes the opportunity to avail themselves of various government reservation facilities. Those who are able to attain higher education often secure jobs in different educational and administrative institutions and have migrated to the cities. Some families have built houses in the Doon Valley, although they continue to reside in their rural settlements. Unlike other parts of the Uttarakhand Himalaya, the rate of outmigration from Jaunsar-Bawar and Rawain is relatively low. It has been observed that agricultural output is higher here compared to other parts of the Himalaya, which is why the farming community continues to practice agriculture.

3.19 Occupation

The primary occupation of the people in the region is agriculture, with both men and women, including children, working in the agricultural fields. They grow various types of food grains, millets, fruits, vegetables, and spices using traditional farming methods. This is subsistence agriculture, where the products are mainly consumed domestically. Recently, the cultivation of fruits, especially apples, has been on the rise. Alongside agriculture, livestock rearing is an allied practice, with people raising cows, goats, sheep, lambs, horses, and oxen. During the summer months, farmers move their animals to alpine pasturelands, where they stay for about four months for grazing. Selling milk and dairy products at local markets provides a source of livelihood. Due to the high number of goats and sheep, wool is extracted, and the community makes woolen clothes (Fig. 3.21), which are essential during the cold winter months. These woolen garments are part of the unique traditional attire of the people. A few individuals are involved in trade and artisan activities. The community also practices several rites that are unique to them, although some rites overlap with those practiced by Garhwalis.

Fig. 3.21 Elderly men are weaving woolen thread in the community place of Osla village. Photo By author

3.20 Tourism as a Source of Income

The Tons River basin offers numerous destinations that are well-suited for tourism development. One such destination is the Supin River, which flows through the Naro Phu range, transforming into the serene Buransh (rhododendron) range. This area is home to rare wildlife, including the Himalayan Thar and the elusive snow leopards, making it a prime location for nature and wildlife enthusiasts. Kalsi, another important site, serves as a trail leading to Hari Ki Dun and is rich in ancient history. It was once ruled by the Emperor Ashoka, as evidenced by the Ashokan edicts carved in stone. Haripur Danda, a hilltop Jaunsari village, is home to the Haripur Temple, adding to the region's cultural significance. On the other side of the Tons River, in Himachal Pradesh, the Rodu settlement offers yet another beautiful destination for visitors, providing a picturesque experience for those seeking to explore the natural beauty and cultural heritage of the area.

Trekking and river rafting are two prominent adventure tourism activities in the Tons River basin. Among the most popular trekking destinations are Hari Ki Dun and Kedar Kantha. These treks attract numerous adventurers each year who camp with tents along the way. Both routes are challenging, with Hari Ki Dun being particularly tough due to its long distance. In addition to trekking, river rafting in the Tons River near Mori town is a popular adventure activity, especially during the winter season. This thrilling experience could significantly promote adventure tourism in the region, offering both locals and visitors an opportunity to engage with the river's powerful currents and scenic surroundings.

Jaunsari culture is distinct, marked by its unique folk deities, fairs, and festivals, which set it apart from other regions in the Himalayas. This creates a strong potential for cultural tourism in the Tons Valley. Two major cultural tourism destinations are Lakhamandal and Hanol. Lakhamandal is deeply connected to the history of the Pandavas, and its temple, dedicated to Lord Shiva and Goddess Parvati, is a significant religious site. In Hanol, the Mahasu Devta temple, situated on the left bank of the Tons River, is a major cultural center. Mahasu Devta, a folk deity believed to be an incarnation of Shiva, draws numerous pilgrims. The valley is also home to other revered folk deities such as Someshwar and Karna. Fairs and festivals are regularly held at these temples, with the cultural procession of Lord Someshwar being a key attraction for visitors. These cultural elements combine to make the Tons Valley an ideal location for cultural tourism, offering travelers a chance to explore its rich traditions and religious heritage.

Homestay tourism is becoming an increasingly important livelihood option in the Tons River basin, driven by the growing number of pilgrims and tourists visiting the area (Sati and Banerjee 2024). Due to the lack of proper infrastructure in the region, particularly along the Tons River, many households, especially those located on the roadside, have turned to homestay tourism. These households provide accommodations, along with local food and beverages, to visitors. This not only allows tourists to experience the region's authentic culture and cuisine but also offers substantial income to the local families, making homestays a significant economic activity in the area. The rise of this tourism model highlights the potential of local communities in sustainable tourism development.

3.21 Conclusions

The Tons River basin is majestic in all aspects—natural, cultural, social, and spiritual. The landscape is spectacular, featuring various physiographic features such as the snow-capped high Himalayas, alpine grasslands, Middle Himalayas, river valleys, the Lesser Himalayas, and Shivalik ranges. The "U"-shaped and "V"-shaped valleys, rapids, rivulets, streams, waterfalls, river meanders, gorges, terraces, and broad valleys create panoramic landscapes that attract tourist interest. The cultural landscape is rich, unique, and diverse. People perform distinct cultural practices, customs, rituals, and fairs and festivals, which are celebrated according to the changing seasons. People worship folk deities on different occasions and offer sacrifices of goats to appease them. Society is woven together by different castes and creeds (Sati, 2022). They are economically underdeveloped, and therefore, they have been granted ST status to aid their social and economic upliftment. The economy is mainly based on traditional agriculture, while other allied activities such as livestock farming and homestay tourism have the potential to contribute to the region's economic development. Trekking to Hari Ki Dun and Kedar Kantha is a lifetime achievement, although it is very tough. The entire region is majestic and divine. While Hari Ki Dun is renowned as the Valley of God, the entire region is considered an abode of Shiva and Shakti. This study provides a comprehensive picture of Hari Ki Dun and the Tons River basin.

References

Berreman, Gerald D (1963) Review: himalayan polyandry: structure, functioning and culture change. a field-study of Jaunsar-Bawar. Am Anthropol 65(5):1158–1160. https://doi.org/10.1525/aa.1963.65.5.02a00320

Bhardwaj M, Uniyal VP, Sanyal AK, & Singh AP (2012) Butterfly communities along an elevational gradient in the Tons valley, Western Himalayas: Implications of rapid assessment for insect conservation. Journal of Asia-Pacific Entomology, 15(2), 207–217.

Gangopadhyay D, Mukhopadhyay AK (2010) Participatory learning and action for sustainable rural development in India: a case study of Didhsari Village, Uttarkashi, Uttarakhand. J Rural Dev 29(2):119–139

Gazetteer of India (1909) Chakrata Tahsil & Town The Imperial Gazetteer of India, 1909, v. 10, p. 125

Jain SP, Puri HS (1984) Ethnomedicinal plants of Jaunsar-Bawar hills, Uttar Pradesh, India. J Ethnopharmacol 12(2):213–222. https://doi.org/10.1016/0378-8741(84)90049-7

Kala CP, Dhyani PP, Sajwan BS (2014) Developing the medicinal plants sector in northern India: challenges and opportunities. Appl Ecol Environ Sci 3(1):38–44

Kumar M (2022) Phytosociological and socioeconomic studies of traditional agroforestry systems in Naugaon block of Uttarkashi district, Uttarakhand (master's thesis). G. B. Pant University of Agriculture and Technology. https://krishikosh.egranth.ac.in/items/70ef503b-5068-46d8-9cff-3381be9bc313

Majumdar DN (1962) Himalayan polyandry: structure, functioning and culture change; a field-study of Jaunsar-Bawar. Asia Publishing House, New York

Papnai GS, Negi SS, Rawat LS (2017) Performance of improved cultivars of French bean, garden pea and okra in Uttarkashi district of Uttarakhand. Indian J Hill Farming 30(2):167–169

Rana TS, Datt B (1997) Ethnobotanical observation among Jaunsaris of Jaunsar-Bawar, Dehra Dun. Int J Pharmacol 35:371–374. https://doi.org/10.1080/09251619708951285

Rawat V (2023) Study on traditional agroforestry systems in Mori block of Uttarkashi district of Uttarakhand (master's thesis). G. B. Pant University of Agriculture and Technology. https://krishikosh.egranth.ac.in/items/293e3881-aa63-4f60-a9a0-ae5a1d61c24e

Rawat A (2018) Functional change of rural settlements in the high Himalayan region: a case study of Sankari, Uttarkashi. Res J Hum Soc Sci 6(1):6–16

Sathyakumar S (1994) Habitat ecology of major ungulates in Kedarnath Musk Deer Sanctuary, Western Himalaya. Ph.D. Thesis, Saurashtra University, Rajkot. 242 pp.

Sati VP (2025) Sustainable homestay tourism for enhancing rural livelihood in the Uttarakhand Himalaya: a study of the tons River Basin. BIO Web Conf 151:03001. https://doi.org/10.1051/bioconf/202515103001

Sati VP, Banerjee S (2024) Enhancing rural livelihoods through agro-biodiversity conservation, sustainable agriculture, and homestay tourism in the tons River Basin, Garhwal Himalaya. Malaysian Journal of Sustainable Agriculture 8(4):41–50

Sati VP (2023a) Cultural processions in the central Himalaya: historical and spiritual significance. J Cult Stud 16(4):50–64. https://ojs.unud.ac.id/index.php/ecs/

Sati VP (2023b) Natural and cultural diversity in the Himalaya. Springer, Cham

Sati VP (2023c) Pilgrimage to the Himalaya: historical perspectives and present scenario. J Tour Sustain 6(2):39–48. https://doi.org/10.5281/zenodo.8044493

Sati VP (2023d) Pilgrimage tourism in Uttarakhand Himalaya: pilgrims' inflows and trends. J Multidiscip Acad Tour 8(2):109–117. https://doi.org/10.31822/jomat.2023-8-2-109

Sati VP (2023e) Sustainable forest management in the Himalaya. Springer, Cham

Sati VP (2022) Uttarakhand: society, culture, and pilgrimages. Today and Tomorrow, Printers and Publishers, New Delhi

Sati VP (2021) The Ganges: cultural, economic, and environmental significance. Springer, Cham

Sati VP (2018) Carrying capacity analysis and destination development: a case study of Gangotri tourists/pilgrims' circuit in the Himalaya. APJTR 23(3):312–322. https://doi.org/10.1080/10941665.2018.1433220

Singh DS (ed) (2018) The Indian rivers : scientific and socio-economic aspects. Springer, Singapore, p 416

Uttar Pradesh District Gazetteers (2021) Uttarkashi. Indian culture. Accessed 23 May 2021

Chapter 4
The Legends of Kedarkhand: A Journey Through the Himalayan Abode

Contents

Abstract The western part of the Uttarakhand Himalaya is known as Kedarkhand. It holds immense ancient and cultural significance, as it is home to the world-famous four pilgrimages (Char Dham). This region is also the land of the Panch Badris, Panch Kedars, and Panch Prayags. This chapter elaborates on the natural and cultural landscapes along four significant routes within Kedarkhand. The first route extends from Haridwar to Badrinath and Kedarnath; the second from Rishikesh to Gangotri and Yamunotri; the third from Karnprayag to Homekund, Sutol, and Nandprayag; and the fourth from Haridwar to Kotdwar, Pauri, and Srinagar. The cultures, economies, and environments along each of these routes are discussed in detail, supported by relevant illustrations.

Keywords Badrinath · Kedarnath · Gangotri · Yamunotri · Panch Prayags · Kedarkhand

V. P. Sati, *Exploring the Mystical Himalaya*, Springer Geography,
https://doi.org/10.1007/978-3-032-14275-7_4

4.1 Introduction

Kedarkhand is referred to as the abode of Shiva one among the most sacred regions of the Himalaya, which has been a divine space in Hinduism. In the Hindu texts Puranas, Kedarkhand is presented not only as a geographical region but also as metaphysical space, a spiritual entity and an area where gods and sages communed with the divine (Bharati 1965). The Skanda Purana (1968), one of the largest and most ancient texts, describes Kedarkhand a place of worshiping Shiva and Shakti and which has a deep spiritual significance. It describes the entire region as blessed by Lord Shiva, particularly the sacredness of the Kedarnath Temple, one of the twelve Jyotirlingas. Legend says that Shiva appeared as a bull (Nandi) and took refuge in Kedarnath escaping from the Pandavas. Later on, Pandavas sought blessing from Shiva to absolve their sins from Mahabharata War. Pandavas built a temple where Lord Shiva appeared, which is now the highland pilgrimage Kedarnath, situated on the bank of the Mandakini River. Pilgrim visit Kedarnath to wash their sins and to attain Moksha (salvation).

The Vayu Purana (1980) states that Kedarnath is considered as pilgrimage destination par excellence. It is mentioned that anyone who undertakes a pilgrimage to Kedarnath and bathes in the holy rivers, achieve Moksha from the cycle of birth and death (Chakrabarty 2015). In the Purana, it is mentioned that the ancient sages like Narada and Vyasa performed penance in Kedarkhand. Garuda Purana (1908) states that Kedarnath is a place where gods aspire to visit, as it is home to the sacred Jyotirlinga. It is said to represent the element of earth (Prithivi). Its location in the towering Himalaya reinforces the idea of transcendence from the material to the spiritual. The Shiva-Shakti connection here also illustrates the balance required in the universe (Shastri 1983). The Kedarkhand represent for Hindu scared geography combining physical geography, mythology, and spiritual quest (McDaniel 2004).

The Tantra has high essence in Kedarkhand (Brooks 1990; Kinsley 1997). It is often misunderstood, whereas it has high spiritual significance through it, divine experience can be attained. Tantra comprises both the material and spiritual worlds as divine manifestation. Bhakti is another tradition, which emphasizes love and devotion to a personal deity (Sivananda 2003). In Kedarkhand, pilgrims undertake the visit of Kedarnath with pronouncing Har-Har Mahadev as they ascend to mountain. This is an act of devotion, which connects to the divine power. Aarti is another devotional act, performed at morning and evening to appease deity. Tantra and Bhakti are performed with different ways and can be seen at all the pilgrimages in Kedarkhand; however, there essence is similar as to connect the divine power.

Kedarkhand is known as the Garhwal region in the recent period (Fig. 4.1). It comprises of the two different socio-cultural realms—Garhwal and Jaunsar-Bawar, and Rawain realms. Garhwal combines two words—Garh and Wal as there were 52 Garhs (fortress) before independence. It is separated from the Manaskhand through a longitudinal line from north to south. The line divides the Chamoli district from Pithoragarh, Bageshwar, and Almora; Pauri from Almora and Nainital; and Haridwar from USN. Kedarkhand had many sub-regions of their geographical, economic,

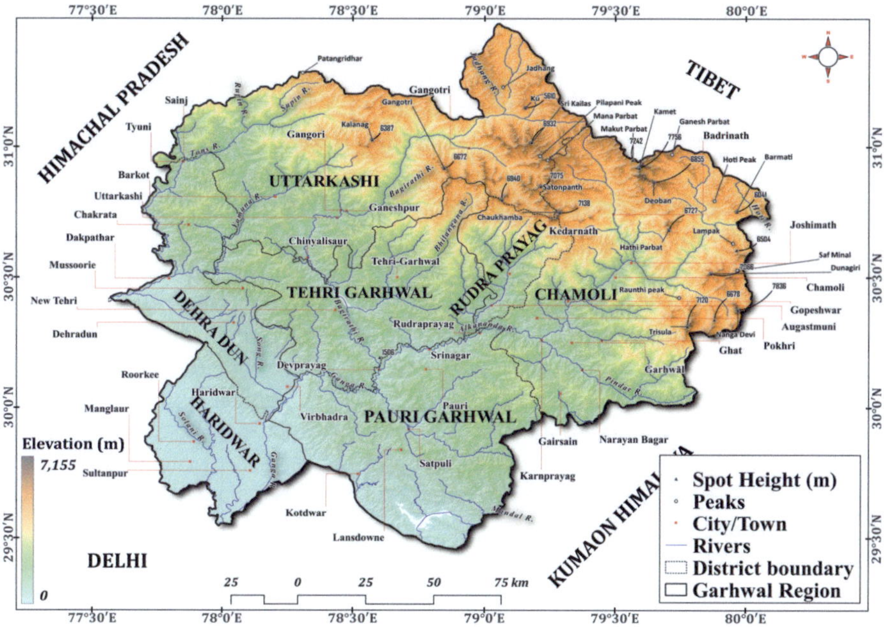

Fig. 4.1 Location map of Kedarkhand

social, and cultural importance. Some of them were Painkhanda, Dasholi, Chandpur, Nagpur, Badhan, Gangapaar, Salan, and Raath. Now these sub-regions are nomenclature as different names. This region has spectacular natural, cultural, social, and forest landscapes and has been important for the ancient period. The author had some important visits of the rich cultural destination—highlands and river valleys. In this chapter, he illustrated some of the important natural and cultural landscapes of Kedarkhand.

4.2 Economy and Environment

The economy of Kedarkhand is underdeveloped and primarily based on the cultivation of traditional cereal crops (Sati 2012; Gopal et al. 2025). Livestock farming is an allied activity and plays a significant role in sustaining livelihoods. Approximately 70% of the population practices agriculture and livestock farming. Dairy farming has significant potential for economic development (Chanyal et al. 2019; Nagar et al. 2025). However, the output from biomass-based agriculture is insignificant, and as a result, a large number of people live below the poverty line. Recently, the impact of climate change on agriculture has been substantial, leading to a decline in agricultural productivity (Anshu and Pandey 2024).

The level of education is substantial, and those who are highly educated have out-migrated to other parts of the country in search of jobs. Recently, migration has become a serious problem in rural areas, leading to land abandonment and low crop production and productivity. Infrastructural facilities are poor, including transportation, communication, and market access, which further hinder development processes (Kakkar and Mishra 2025).

Tourism, as an emerging sector of economic development, has very high potential, as Kedarkhand is rich in cultural, natural, and adventure destinations (Singh et al. 2024; Chand and Mishra 2024). It receives hundreds of thousands of tourists and pilgrims every year, enhancing the income of rural people and boosting the economy of the region and the state as a whole. It also plays a significant role in improving the State Gross Domestic Product (SGDP), and its share is high in Kedarkhand. Community participation in tourism development, along with government support, may lead to sustainable economic development (Verma et al. 2024). The Garhwal region has rich food and beverage traditions, with local cuisines served to tourists and pilgrims. These can be further promoted as a major attraction in tourism destinations (Lingwal et al. 2022).

Environmental quality throughout Kedarkhand is generally high and healthy. The climate ranges from subtropical to temperate, cold, and frigid conditions. However, natural disasters are highly frequent, particularly during the rainy season. Each year, multiple disasters occur, resulting in significant loss of life and property (Jain et al. 2017). The major disasters in the region include cloudburst-triggered debris flows, flash floods, landslides, and mass movements (Rana et al. 2022; Dobhal et al. 2013). Additionally, avalanches, Glacial Lake Outburst Floods (GLOFs), and glacier bursts are other prominent disasters (Shrestha et al. 2010).

4.3 Natural and Cultural Landscapes

4.3.1 Haridwar to Badrinath and Kedarnath

This is the most important natural and cultural landscape of Kedarkhand, started from Haridwar (200 m). Haridwar (Located in Haridwar district) is renowned as the Gateway to God and the spiritual capital of Kedarkhand. It has a long history dated back to the King of the Himalaya—Daksha Prajapati, Goddess Sati, and Lord Shiva. After humiliated by father Daksha Prajapati in the Great Yagna, which was attended by Brahma, Vishnu, all gods, goddesses, sages, and saints, Goddess Sati jumped into the burning fire pod and died. This led to the establishment of the Shakti Peeths in the Himalayan Region. Haridwar is situated on the banks of the holy Ganga River, having two famous Shakti Peeths—Chandi Devi and Mansa Devi, and a Ganga temple, where evening Aarti is performed every day. Haridwar is also known for a holy dip in the Ganga. Ganga Dussehra and Kumbh are the two major fairs celebrated in Haridwar. Haridwar has a funeral ground on the bank of the Ganga, known as Moksha Dham. Taking a holy dip and performing funeral in the Ganga River is believed to liberate humans from the cycle of birth and death. Narayani Shila is a rich cultural place, where people perform rituals to ascent their forefathers to Swarga. A 24 km saffron belt goes to Rishikesh along the Ganga River. Rishikesh is another popular river valley pilgrimage, renowned as the Yoga capital of the world. It has many cultural places and some of them are Triveni, Parmarth Niketan, and Bharat Sadhu Samaj. Rishikesh, spanning in the three districts of Dehradun, Pauri, and Tehri, is another name of Lord Vishnu and it a place where the Ganga enters into the plain region from its undulating mountainous region. Both Haridwar and Rishikesh are the popular centers of natural and cultural tourism and exodus people from the Indian sub-continent and abroad visit throughout year. River rafting is practiced in the Ganga River in Rishikesh (Fig. 4.2). After Rishikesh, a narrow "V" Shaped valley of the Ganga starts and goes continue till the place of its origine. The road goes through the right bank of the Ganga River until Vyasi, which is 30 km long. The word Vyasi is derived from the name of famous saint (Muni) Vyas. Delicious food is served as breakfast and lunch in Vyasi.

The road continues goes through the bank of the Ganga River and after passing Kodiyala, a famous river rafting side and a hill resorts, the road goes upwards up to Sakinidhaar, about 30 km. Then road goes downwards passing Tota Ghati, a high landslide prone area and after 30 km, it reaches to Dev Prayag, a holy town, located in two districts of Tehri and Pauri. The Alaknanda River joins the Bhagirathi River at Dev Prayag and both rivers after joining known as the Ganga, the holiest river of the world. A famous Raghunath Temple is located at the confluence (Sangam) of both rivers. The Sangam at Dev Prayag is as important as Haridwar and Rishikesh

Fig. 4.2 The Ganga River is flowing in Rishikesh, the Gateway to the Himalaya; Photo by author

and the tourists and pilgrims take a holy dip at the Sangam on the auspicious occasions and throughout the year. The Pandas of Dev Prayag are very famous for performing rituals and their working areas in complete north and central India. The road bifurcates from Dev Prayag, one goes to New Tehri and other goes to Srinagar Garhwal, which is located on the way to the highland pilgrimage Badrinath.

The road goes through the bank of the Alaknanda River up to Kirti Nagar. It makes the beautiful landscape along with arable river terraces. A bridge crossing the Alaknanda River at Kirti Nagar and the road again goes through the bank of the Alaknanda River. A beautiful city Srinagar Garhwal is situated on the banks of the Alaknanda River, which right bank falls in the Tehri district and left bank is the Pauri district. Srinagar is the cultural city of Garhwal, popularly known for quality education. Many quality educational institutions—secondary and higher are imparting education to the students of the region. One of the old and renowned university is the HNB Garhwal University. Srinagar city has mixed Garhwali and Kumaoni culture. Educational institutions have integrated the rich culture of Garhwal and Kumaon. The city is renowned for its folklore—songs and dances. The Alaknanda River makes a spectacular meander and a stretch river valley. A 400 mw hydropower project is constructed on the Alaknanda River at Srinagar (Fig. 4.3). Some important Shiva temples are located in the city such as Ghasiya Mahadev and Kamleshwar Mahadev. Shrikot was a village, located in the river terrace of the Alaknanda, is now a part of Srinagar city, where a 200 bedded base hospital is located, which is serving medical services to a vast population of the Garhwal region.

Approximately 20 km, Dhari Devi Shakti Peeth is situated on the Alaknanda River. It was situated on the bank of the Alaknanda River on a big stone. Later on,

Fig. 4.3 A reservoir of hydropower project near Dhari Devi Temple, Shaktipeeth. Photo By author

the stone was submerged into the reservoir of Srinagar Dam and their a temple constructed on the Alaknanda River on the pillars. This is highly revered temple were hundreds of thousands of pilgrims visit every day. It makes spectacular river valley landscape. During the monsoon season, when the water of reservoir rises, the temple sink almost half in the water. Although the Kedarkhand is highly fragile, vulnerable to landslide hazards, yet, this area is the most fragile and vulnerable. A centuries old landslide "Kaliyasaud" is very active zone for landslide. Every year, it causes many actualities. With many ups and down, and curvy roads, it goes to Rudraprayag along the bank of the Alaknanda. Rudraprayag is a town and a district headquarters. Rudra is the manifestation of Lord Shiva and Prayag is Sangam of the Alaknanda and the Mandakini Rivers. A temple of Rudra (Shiva) is situated on the Sangam where devotees worship. Rudraprayag is located in the prime location because the road bifurcates leading to two world-famous pilgrimages—Kedarnath and Badrinath. The Alaknanda River flows through a deep valley making gorge at Koteshwar Mahadev Temple near Rudraprayag (Fig. 4.4).

The road continues goes to Karnprayag along the Alaknanda River. After a distance of 30 km, a beautiful river valley town Karnprayag is located where two rivers the Alaknanda and Pindar meet. Karnprayag is a cultural and educational town. The famous Karna and Lord Krishna Temples are located on the stone, on the bank of both rivers. On the Sangam between the two rivers, the famous folk deity Uma Devi Temple is situated. Uma is the manifestation of Nanda Devi, an adorable deity of the region. On the Sankranti of Vaisakh, Vaisakhi festival is celebrated on the Sangam, where the people from surrounding areas visit and celebrate the fair. This festival is

Fig. 4.4 The Alaknanda River is making gorge at Koteshwar Mahadev Temple, right side of it near Rudraprayag. Photo By author

also called Bikhoti. The devotees take a holy dip at the Sangam. Karnprayag has a range of hotels and the local food and beverages served to tourists are delicious. The proposed capital of Uttarakhand State is located about 50 km from Karnprayag. The road goes continues to Chamoli, district headquarters of the Chamoli district along the Alaknanda River via Langasu, Nand Prayag, and many other rural settlements, located on the river terraces with plenty of arable land. High crop diversity is found in these settlements. Nand Prayag is famous for meeting to the Nandakini River with the Alaknanda River. It makes a wide and stretch valley at the Sangam, where Lord Shiva and Goddess Shakti temples are located. Along with practicing agriculture and rearing animals, the people also serve tourists and pilgrims providing them accommodation and local food and beverages. A small town Gopeshwar is located 8 km from Chamoli town, which is renowned for the temple of Gopinath, a manifestation of Lord Shiva.

From Chamoli town, the road first goes through the narrow valley making several gorges and then the road turns upwards with steep slopes up to Joshimath (now Jyotirmath). The entire mountainous road is prone to road accident. Pipalkoti is a rural settlement, located before Jyotirmath is known for a trekking of 2 km to Kalpeswhar, a pilgrimage site and a form of Lord Shiva. Jyotirmath is the winter abode of Lord Vishnu. During the six month of winter, when Badrinath, the summer abode of Lord Vishnu receives heavy snow. The idol of Lord Vishnu is brought to Jyotirmath. Jyotirmath (2000 m) is also known for the famous and beautiful temple of Narsimha. It is a hill town, pleasant for the summer and it receives heavy snow during the winter. A road bifurcates to Niti Pass along the course of the Dhauli Ganga, the border area with Tibet. The famous Auli, a place for winter skiing, is located 13 km upwards of Jyotirmath. Because of the border area, Jyotirmath has a huge army camp. This area is renowned for the cultivation of potato, apple, and Rajma.

Vishnu Prayag is situated on the meeting points of the Alaknanda and Dhauli Ganga, 10 km downslope from Joshimath. This road leads to the highland pilgrimage Badrinath. The distance between Joshimath and Badrinath is about 60 km and the entire area is highly fragile and the most vulnerable to landslides. There are many evergreen snow-fed streams joins the Alaknanda from both sides. These streams become violent during the monsoon season and cause for flash floods and landslides. Pagalnala in Lambagar is the most worst place often remain closed due to landslides and flash floods. Govind Ghat a pilgrimage for the Sikhs, and known for its Gurudwara, is located just in the middle of Joshimath and Badrinath. The bridal road leads to Ghanghariya, the Valley of Flowers National Parks, and Hemkund Sahib, the famous Sikh highland pilgrimage, located more than 3000 m. A Vishnu Prayag hydropower project of 400 mw is constructed on the Alaknanda on the way to Badrinath. The road goes upwards through meandering roads for about 20 km and reaches Badrinath. Badrinath (3300 m) is located on the banks of the Alaknanda, which is known as the Vishnu Ganga in Badrinath. Badrinath is an abode of Lord Vishnu, surrounded by Nar and Narayan Mountains, a place where Vedas were written by Ved Vyas. The statue of Lord Vishnu is laying down in the

Garbhagriha and pilgrims worship him. A long que goes to the bank of the Vishnu Ganga and people wait for Darshanam. Morning and evening Aarti is performed.

Lord Vishnu has five manifestations and they are renowned as Panch Badris. Four Badris—Badri Vishal, Bridha Badri, Yogdhyana Badri, and Bhavishya Badri, are located in the Upper Alaknanda Basin and the fifth one, Adi Badri is located in the Pindar basin, close to Karnprayag. The main priest of the temple is Kerala's Namboodiri, who serves the temple. Others are the Brahmins of Garhwal. This shows the national integration. The most spiritual place for the Hindus, Badrinath is a famous highland pilgrimage, one amongst the Four Dhams of India. Rishi Narada performed penance here, as a Narada Kund is located close to the temple of Badrinath. On the right bank of the Vishnu Ganga, many hot water ponds (Geysers) are located where the pilgrims take holy dips. During the six month of winter, Badrinath Temple remains closed.

A 3 km from Badrinath, Mana village is located, which is now known as the first village of India. Mana village is located on the banks of Alaknanda and Saraswati. Both rivers meet here and make Sangam. A famous natural stone bridge, known as Bhim Pul, is located on the Saraswati River, a major attraction of the pilgrims and nature levers. After Mana, the land is baren, stony, and snow-capped. A few kilometer distance, an international border with Tibet starts. The Tibet border was opened before the Chinses war of 1962 and India had good relationship with Tibetans. A 30 km trekking goes to Swargarohini from Badrinath. Beneath Swargarohini, a glacial-fed lake Satopanth is situated, which remains frozen during the winter. This is the toughest trekking, on the way of which is a biggest waterfall "Vasudhara." The Hindu scripture says that Pandavas along with Dropadi trekked to Swargarohini before leaving their earthly abode. The Badrinath valley is India's most important natural and cultural destination (Fig. 4.5). Full of sanctity, the divine waves blow in the valley. A famous Badri Van is found in surrounding of Badrinath Dam. On the way to Satopanth, Brahma Kamal (God's flowers grow during summer).

The pilgrims come back to Chamoli town via same route and turns to Kedarnath, another highland pilgrimage. The road bifurcates to Gopeshwar town and Mandal, a rural settlement. The Mandal valley is spectacular and known for its cultural and natural landscapes. It is famous for the temples of Rishi Atri and Mata Anusuya, which is located approximately 5 km from Mandal. Another trek of 17 km leads to Rudranath highland pilgrimage, one of the Panch Kedars. A world-famous Nanda Devi Biosphere Reserve starts from Mandal and the road to Kedarnath goes through the core of it. It is more than 50 km road up to Chopta, a small service center, known as the Switzerland of Uttarakhand. A 3 km trek goes to Tungnath (another manifestation of Lord Shiva) from Chopta, above which the Chandra Shilla, a confluence of spectacular natural and cultural landscapes, is located. The road goes continues to Ukhimath via Pothibasa and Dugglebita, the two rural settlements, known for their beautiful rural settlements. Ukhimath is a holy place, which is famous for the Omkareshwar Temple, the abode of Lord Shiva. It is a winter abode of Lords Kedarnath and Madhyamaheshwar. When the highland sacred places—Kedarnath and Madhyamaheshwar receive heavy snowfall, the idols of deities are brought to Omkareshwar Temple of Ukhimath and it remains here for the 6 months. From

Fig. 4.5 Brahma Kamal (God's flower) grows on the way to Satopanth lake, 30 km from Badrinath Dham

Ukhimath, a trekking of more than 20 km goes to Madhyamaheshwar, a highland pilgrimage and one of the Panch Kedars. The road goes downwards about 10 km, crosses the Mandakini River, and then goes upwards to Gupta Kashi. On the Bank of the Mandakini River, an important cultural place Kund is situated, which is very ancient, having historical importance. On the confluence of the Kali Ganga and Saraswati Ganga, a famous Shakti Peeth "Kalimath" is located, which is believed to be a place where, Kalidas, the famous poet and writer stayed during his old-age and he composed the famous epic (Maha Kavya) "Meghdoot."

Gupta Kashi is a town and service center where the famous temple of Lord Shiva lies. The road goes continues to Gaurikund from which, a trekking of 16 km to Kedarnath starts. Gauri is believed to be an incarnation of Parvati, wife of Lord Shiva. Gouri Kund, located on the bank of the Mandakini River, is a hot spring (geyser), where pilgrims take holy bath before trekking to Kedarnath. Before Gaurikund, and famous Prayag "Son Prayag" lies where the Son Ganga joins the Mandakini River. After which, a road goes to Triyuginarayan, famous for a Shiva Temple. It is believed that Lord Shiva married with Goddess Parvati in this place. There is a fire pod, which has been burning from the day of the marriage of Shiva and Shakti, as it is believed. Recently, the State Government has declared it as a wedding destination.

Along the left bank of the Mandakini River, a trekking route of 14 km traversed to Kedarnath through Rambada before 2013. The entire route, including Kedarnath pilgrimage, was washed away due to cloudburst-triggered debris flows and flash floods, after which, a new trekking route of 16 km was constructed to Kedarnath along the right bank of the Mandakini River. This trekking route is toughest route and during the pilgrimage to Kedarnath, occurrence of landslides is common. However, the entire route is spectacular, a divine path. Pilgrims visit to Kedarnath from all over the world. After a long and tired trekking through ups and downs and various forest landscapes, pilgrims reach the Kedarnath pilgrimage. Kedarnath is a notified urban center, a summer dwelling of Lord Shiva, priests, and the service providers. During winter, they all come down with their animals. Kedarnath lies on the Bank of Mandakini River (Fig. 4.6). There are four other small streams flowing through Kedarnath. The surrounding mountain peaks of Kedarnath remains snow-capped throughout the year. Above Kedarnath, a glacier-fed perennial lake "Vasuki" lies, which is feeding to the Mandakini River. All Panch Kedars are located in the alpine pasturelands above 3000 m just below the snow-capped mountains of the Himalaya. They are beautiful landscapes on the earth showing their natural and cultural significance. Divine waves blow in these highland pilgrimages. After visiting Kedarnath, the pilgrims go back to Haridwar via Rudraprayag and so on.

The natural and cultural landscapes of entire route are spectacular with rich diversity. It is the home for Panch Badris, Panch Kedars, and Panch Prayags and these cultural destinations are world-famous although there are more than 17 Prayags and many pilgrimages and scared places. Similarly, the river valleys, terraces, cascades, rapids, waterfall, river meanders, gorges, V-shaped valleys, highly elevated mountain peaks, U-shaped valley, moraines, alpine pasturelands, rich biodiversity in fauna and flora, and several human races, make the natural and cultural landscapes spectacular. The entire route is visited by hundreds of thousands of tourists and pilgrims every day.

Fig. 4.6 Kedarnath highland pilgrimage before June 2013. Photo By author

4.3.2 Rishikesh to Gangotri and Yamunotri

Rishikesh to Gangotri and Yamunotri route characterizes both natural and cultural landscapes and there are many trails to visit these highland pilgrimages and other important destinations of natural and cultural importance (Sati 2021). The popular trail is from Rishikesh to Gangotri via Narendra Nagar, Chamba, and Uttarkashi. A road bifurcates from Rishikesh to Narendra Nagar and goes to Chamba. It is a mountainous road goes through hills and valleys. Narendra Nagar, situated on the mountain slope at an altitude of about 1326 m (average) was the capital of Tehri King before India's independence. A palace of Tehri King is still existed, which has been converted into a Hotel, where dignitaries from all part of India stay throughout the year. Narendra Nagar is also a cultural center. Nanda Devi Raajjat, which comes every 12 years, begins from Narendra Nagar. The opening and closing dates of Badrinath Dham is decided from Narendra Nagar. Close to Narendra Nagar, the famous Shakti Peeth Kunjapuri (1676 m) is situated on the top of the mountain and in a serene environment. It is a trek of around 3 km to reach the premises of the temple from the main road. The temple remains open throughout the year for the devotees, whereas during Navratris of Chaitra and Ashvin, fairs are celebrated. The devotees from different part of the state join the festival and worship the Shakti in the temple premises.

The road continues goes to Chamba, a hill town and summer resort. A road bifurcates from Chamba to Dhanaulti via Kaddu Khal. Kaddu Khal is a place, from a 2 km trek goes to Surkanda Devi Temple. It is a famous Shakti Peeth in Uttarakhand. The trek goes through rough and rugged terrain with steep slope. Some dense patches of Deodar trees are found on the way. Monkey and languor are also found abundant. The temple is very ancient while recently reconstructed with beautiful architect. The premises of temple are situated on the top of the hills, quite, calm, and spectacular. A long belt of the snow clad Himalaya can be seen from the temple premises. Devotees worship the deity and take a walk of the temple premises. Immediately reaching the temple premises, the devotees find all well with fresh body and mind. The state government has constructed ropeways of 2 km from Kaddu Khal to the temple premises recently, which is well operational. During Navratris of Chaitra and Ashvin, fairs are celebrated in temple premises. This region is famous for the cultivation of citrus fruits mainly orange and lemon—big and small. Besides, pulses and seasonal vegetables are grown largely.

After visiting Surkanda, the pilgrims go back to Chamba. The road bifurcates to Chandrabadani via Badshahithaul and New Tehri and finally reaches to Kandi Khal. An 8 km trek goes to Chandrabadani temple from Kandi Khal. Recently, a road has been constructed up to the temple premises. Chandrabadani is a Shakti Peeth, remains open throughout the year for performing Puja and large number of pilgrims visit the temple. On the various occasions, mainly during Navratris of Chaitra and Ashvin, nine days fairs and festivals are celebrated. After a pilgrimage to Chandrabadani, the pilgrims proceed back to Chamba and then then toward Uttarkashi along the valley of the Bhagirathi River. A 76 km long reservoir is constructed in the Bhagirathi and Bhilangana Rivers, which is known as the Tehri high dam. The road goes along the left bank of the Tehri high dam reservoir up to Chinyalisaud, a village and an army camp. An airport is located here. At Chinyalisaud, The Bhagirathi River makes a wide and stretch valley. The next destination is Uttarkashi town.

Uttarkashi town is a district headquarters of the Uttarkashi district, a border district of Uttarakhand, make the boundary with Tibet in the north. It has geo-strategic location and the district is very important in its natural and cultural diversity. The famous Govind wildlife sanctuary and Gangotri National Parks are located in this district. In addition, the two world- famous highland pilgrimages—Gangorti and Yamunotri are also located in this district. In Uttarkashi town, a Pala-Maneri dam phase II is constructed on the Bhagirathi River.

Uttarkashi town is renowned as the Kashi of North. It has quite similarly with Varanasi. For example, Varanasi is situated between the Varuna and Asi Rivers. Uttarkashi is also situated between Varuna and Asi. Baba Vishwanath temple is located in both places (Fig. 4.7). The Ganga Temple and Manikarnika Ghat are located in both places. Evening Aarti of Baba Vishwanath and Ganga is played in Uttarkashi every day. Many fairs and festivals are celebrated in Uttarkashi throughout year. One of the important festivals is Makar Sankranti and it is a 20 days festival. During the pilgrimage season, Uttarkashi remains an important place because the pilgrims visit Gangotri and Yamunotri passing through Uttarkashi.

This region is inhabited by people belonging to four different castes. Brahmins, Rajputs, and Scheduled Castes inhabits in the valley and middle altitudes whereas the Scheduled Tribes inhabit in the higher altitudes, close to the Tibet border. The

Fig. 4.7 Vishwanath Temple in Uttarkashi Town, on the way to Gangotri. Photo By author

Scheduled Tribes are constituted mainly by Jadh Tribes. They are Tibeto-Mongolian race, also known as Marchha and Bhotiya Tribes. Therefore this region has rich cultural fabrics. Jadh Tribes celebrate Johar festival. They live in Nelang and Jadang valleys of Gangotri National Park in the summer and during winter, they migrate to the valley regions and therefore, they hold two dwellings.

Traditional subsistence agriculture is the main occupation and source of livelihood of people. Livestock farming and collection of forest products are subsidiary occupations. Agro-biodiversity is rich as there are more than 50 crop races/cultivars are grown (Sati 2024). Potato, Rajma, pulses, apple, plum, peach, pear, leafy vegetation, and oilseeds are grown (Sati 2004). The agricultural products are pure, free from the chemicals as the farmers practice organic farming. Fruits and pulses are sold in the regional markets.

From Uttarkashi, the route to Gangorti is narrow crossing many waterfalls, rapids, cascades, streams, and rivers. The forest landscape is spectacular on the way. Generally, coniferous forests grow in this region. The entire region remains scow-capped during December, January, and February. The summers are pleasant. Pilgrimage to Gangotri and Yamunotri is practiced during the summer whereas winter tourism is also practiced. However, this region falls under eco-sensitive zone and it is ecologically fragile, whereas rich in biodiversity. On the way, Maneri phase I hydropower project is constructed, which also makes a beautiful landscape. Gangoli village is located on the right bank of the Bhagirathi River is the winter dwellings of the Jadh tribes, although some tribes have permanent dwelling in this village. The road goes continues crossing the Bhagirathi Rivers several times and through the beautiful river valley landscape. The next settlement is Bhatwadi, where a perennial stream meets with the Bhagirathi River and makes a beautiful Surya Prayag. Gangnani is a village, where Rishi Parasnath performed penance. It is located on the north facing slope, therefore, the climate of the village is very cold. A hot spring (geyser) is located in the village. The people worship Nag Devta. The village name Gangnani means the bank of Ganga. During the summer season, the priest goes to the highland areas for the collection of Brahma Kamal (God flowers) and offers them to the Nag Devta.

Sukhi village, lies about 2200 m on a mountain slope, right side of the Bhagirathi River, has a sun facing village. The village is known for the Rishi Sukh Dev, who performed penance here. To reach Sukhi, the road goes upward making many curves and then it goes down to the Bhagirathi valley, crosses the river and then continues goes to Harshil and Dharali. A dense temperate forests of deodar are found along the road and makes beautiful landscape. During the winter season, the entire region remains snow-capped. Harshil is also known as Harishila and Hari Prayag. A Lakshmi Narayan Temple lies on the confluence of the Bhagirathi River and a glacial-fed perennial river. According to a legend, Tulsi cursed Narayana to be a stone here and therefore, the place is called Harishila. Besides the Bhagirathi River, two huge glacier-fed perennial streams flow here. A mixed population of Jadh tribes, Rajputs, and Brahmins live in Harshil. In the past, the people followed their culture and customs separately whereas now, the different castes of people have mixed-up and most of them have now marriage relationship. Apple is the major crop grown here with high quality and quantity. The Jadh people had trade relationship with Tibetan in the past, which has been vanished after Chinese aggression of 1962.

A 3 km after Harshil (Fig. 4.8), two settlements—Dharali on the left bank and Mukhawa (Mukhimath) on the right bank of the Bhagirathi River are situated. Dharali (Fig. 4.9) is both a village and a service center, providing services (food, beverages, and lodging) to pilgrims. Mukhawa is the winter abode of Goddess Ganga (Fig. 4.10), where her idol brought from Gangotri pilgrimage. She has a temple in the village. A hanging bridge on the Bhagirathi River connects Mukhawa with Dharali, on the main road about 1 km upslope. It is a wide and stretch valley, which makes a spectacular landscape. A mixed races people inhabit the two villages. The priests of the temple belong to Nautiyal and Uniyal, the Brahmin clans. The villages are renowned for high quality and quantity of apple.

In between Dharali and Gangotri, Bhaironghati is situated, where the Jahnavi River joins the Bhagirathi River. The Ganga River is also known Jahnavi. Jahnavi is a glacial-fed perennial river. At Bhaironghati, the Jahnavi River makes one of the longest and deepest gorges. Before Bhaironghati, Lanka is situated on the bank of the Bhagirathi River, where Lord Shiva Temple lies and it also make a deep gorge (Fig. 4.11). Both places have cultural heritage and these places are associated with Goddess Ganga and demon king Ravana. No human settlements are located except only a few of government offices. An iron bridge is constructed on the Jahnavi River and the road goes to Gangotri. The road is narrow but the river valley landscape is spectacular. The Gangotri highland pilgrimage is located in the serene environment on the banks of the Bhagirathi River, whereas the temple of Goddess Ganga is located on the right bank. Gangotri is the summer dwellings of the Goddess Ganga and hundreds of thousands of the pilgrims visit. Accommodation facility is limited, whereas pilgrims' inflow is high.

A small number of pilgrims trek to Gaumukh, which is the main source of the Bhagirathi River. It is about 18 km trekking. Gaumukh is a cow-shaped feature and situated on the route of Mount Shivling. It is an alpine grassland, above which is the snow-capped mountains and a mountain peak, called Shivling. The entire valley of Bhagirathi makes a divine path, full of sanctity. After visiting Gaumukh and

(a) (b)

Fig. 4.8 (**a**) Harshil village (**b**) Mukhawa village, winter abode of mother Ganga

Fig. 4.9 Dharali village and service center, behind which is the snow-capped greater Himalaya. Photo By author

Fig. 4.10 Mukhawa, a winter dwelling of Mother Ganga, is located on the upslope of the right bank of the Bhagirathi River. Photo By author

Fig. 4.11 The Bhagirathi River forming a beautiful gorge in Lanka, on the way to Gangotri highland pilgrimage

Gangotri, the pilgrims back to Uttarkashi and then Dharasu. From Dharasu, a road bifurcates to Yamunotri. There is another road goes to Yamunotri from Dehradun and Mussoorie via Nainbag and Naugaon. The last village in this route is Kharsali, which is the winter dwellings of Goddess Yamuna. During winter, when Gangorti Temple remains closed due to heavy snowfall, the idol of folk deity is brought down to Kharsali. Hanuman Chatti and Janki Chatti are two other importance places on the trekking route of Yamunotri. From Kharsali, Yamunotri is located about 16 km and the pilgrims trek this distance to reach the pilgrimage site. It is a tough trek. A ropeway is being constructed from Kharsali to Yamunotri, which will help pilgrims to reach Yamunotri easily.

Yamunotri is among the four highland pilgrimages of Uttarakhand, situated on the bank of the Yamuna River. The Yamuna River is considered as the holy river in India and it has high spiritual importance after the Ganga River. The legend says that taking a holy dip in the Ganga River and taking Achamanam (a purification ritual) from the Yamuna River liberate a person from the cycle of life and death. The name Yamuna is derived from Yama, the God of Death. According to the Hindu mythology, worshiping Goddess Yamuna makes devotees free from the fear of death. Above the Yamunotri Temple, Yamunotri glacier is located, which is the source area of the Yamuna River. The entire area is highly fragile and highly vulnerable to natural disasters.

The entire route of Gangorti and Yamunotri is unique and it has rich natural and cultural diversity. Landscape varies from the river valleys to middle altitudes and the high Himalaya including alpine grasslands and perpetual snow-capped Himalaya, which make a unique landscape. Forest landscape also varies from monsoon deciduous forests to pine, mixed oak, and coniferous and is spectacular. The region has rich soils and abundant water in the form of streams, rivers, and glaciers. It holds a rich cultural diversity. The languages, food habits, culture, customs, rituals, fairs, and festivals vary from the valley regions to the high Himalayan region. In the high Himalaya, the Jadh tribes have different culture and customs.

4.3.3 Karnprayag to Homekund, Sutol, and Nand Prayag

Karnprayag to Homekund is a holy trek. Karnprayag, situated on the confluence of the Alaknanda and Pindar rivers, is a cultural and educational city. It lies close to the Nauti village, which is the home of the folk deity Nanda Devi, an adorable deity. Every 12 year, a cultural procession of Shri Nanda Devi Rajjat with the idol of Goddess Nanda, Khadu (a four horn goat and five and half-year-old), priests, devotees, and pilgrims goes to Homekund from the Nauti Village (Fig. 4.12). It is a 22 day procession, covers 177 km path, and every day, about 10 km distance is covered. The entire path is renowned as "Nanda Path." The path passes through several sub-cultural realms and spectacular natural landscape. The customs, rituals, fairs, and festivals vary from each cultural realm.

The procession starts with the arrival of Khadu in Nauti village from the Kanswan village, situated in the Ata Gad (watershed), a perennial tributary of the Pindar River. Another village, Chandpur Garhi was the capital of the Garhwal kingdom before independence and Kunwars of the Kanswan are the descendants of the King. The village Kanswan is located in proximity of Chandpur Garhi. Close to Chandpur Garhi, the famous Adi Badri Temples are located (Fig. 4.13). The Kunwars of Kanswan along with Khadu leads the cultural procession from Nauti to Homekund. The entire Nanda Path is undulating and zigzagged. The people who are physically and mentally fit can participate in the procession. A medical team checks the health condition of the devotees at "Wan village" accordingly allow them to proceed to the high Himalaya.

(a) (b)

Fig. 4.12 Rajjat cultural procession. (**a**) Pilgrims proceeding to Homekund (highland lake). (**b**) Devotees are waiting for the Rajjat cultural procession

The agricultural landscape is spectacular. Up to Wan village, the last village of the route, there are small patches of terraced fields where food grains, pulses, oilseeds, fruits, and vegetables are grown. Agriculture is the main option of livelihood (Fig. 4.14)

The procession passes through the village from Nauti (1530 m) to Eda Badhadi and Kanswan (1200) down slope in the Chandpur sub-cultural realm. Other villages located on the Nanda Path are Develgarh, Sari, Malethi, Nauna, and Naini villages, between Nauti and Kanswan villages with a night halt at Kanswan village. This procession proceeds to upslopes after Kanswan village to Kandwal Gaon, Chhekuda (1600 m), and Bhagoti villages, crossing Ata Garh and reaching the Srigoor sub-cultural realms. Goddess Nanda is also known as Bhagoti ki Nanda, where a temple is devoted to her. In Bhadra, a fair is organized in the Nanda Temple. It halts at Bhagoti village and then proceeds to down slope crossing Kewer village and the Kewer Gadhera. Kewer Gadhera divides Srigoor and Badhan sub-cultural realms. The rituals and languages vary in these two cultural sub realms. Narayan Bagar is a service center and a village, located in the banks of the Pindar River (Fig. 4.9). A temple of Narayana is located in this village where a fair is organized on Krishna Janmashtami. The procession goes along the road and in the right bank of the Pindar River, passing the villages—Panti, Ming Gadhera, and Harmani, and reached to Kulsari, where it halts for overnight. The next day, it moves to Nandakeshari (1325 m) via Tharali, a service center. It halts in Nandakeshari. Nandakeshari is renowned as a Triveni as three cultures and idols of folk deities—Nanda of Noti, Nanda of Almora, and Chhantoli of Kurur meet here. The next morning it proceeds to Faldiya village and Deval, a service center and Wan village.

Fig. 4.13 Adi Badri Temple (Lord Vishnu), located on the bank of the Ata Garh, a tributary of the Pindar River

Fig. 4.14 Agricultural landscape in Kaub village, along the Pindar River basin

The procession passes through the village from Nauti (1530 m) to Eda Badhadi and Kanswan (1200) down slope in the Chandpur sub-cultural realm. Other villages located on the Nanda Path are Develgarh, Sari, Malethi, Nauna, and Naini villages, between Nauti and Kanswan villages with a night halt at Kanswan village. This procession proceeds to upslopes after Kanswan village to Kandwal Gaon, Chhekuda (1600 m), and Bhagoti villages, crossing Ata Garh and reaching the Srigoor sub-cultural realms. Goddess Nanda is also known as Bhagoti ki Nanda, where a temple is devoted to her. In Bhadra, a fair is organized in the Nanda Temple. It halts at Bhagoti village and then proceeds to down slope crossing Kewer village and the Kewer Gadhera. Kewer Gadhera divides Srigoor and Badhan sub-cultural realms. The rituals and languages vary in these two cultural sub realms. Narayan Bagar is a service center and a village, located in the banks of the Pindar River (Fig. 4.15). A temple of Narayana is located in this village where a fair is organized on Krishna Janmashtami. The procession goes along the road and in the right bank of the Pindar River, passing the villages—Panti, Ming Gadhera, and Harmani, and reached to Kulsari, where it halts for overnight. The next day, it moves to Nandakeshari (1325 m) via Tharali, a service center. It halts in Nandakeshari. Nandakeshari is renowned as a Triveni as three cultures and idols of folk deities—Nanda of Noti, Nanda of Almora, and Chhantoli of Kurur meet here. The next morning it proceeds to Faldiya village and Deval, a service center and Wan village.

Deval is situated on the bank of the Kail River. The procession goes upslopes and continued till Homekund (Fig. 4.16). It halts in Wan village (2444 m), the last village of the Nanda Path. Here, medical checkup of devotees, who want to proceed further in the high Himalaya, is performed. A famous temple of Latu Dev, brother

Fig. 4.15 The Pindar River is flowing near Tharali town of the Chamoli district, a way to Homekund. Photo By author

of Nanda Devi, is located in the village, which is opened for an hour on the day, when Shri Nanda Devi procession halts in the village.

The procession goes through Gairoli Patal (2722 m) and reached to Bedini Bugyal (3200 m). Bedini Bugyal is the alpine pastureland, a unique creation of nature. It has spectacular cultural and natural landscapes. The priests and devotees perform Dev Devi Naach and sing Jagars to appease fold deities. The toughest route starts from Bedini Bugyal. It goes through Shila Samundra (4000 m), Jyuraglidhar (5000 m), Chandmyaghat (4800 m), and finally Homekund (4000 m), where the Khadu is left with idol of Nanda. The same day it back to Nauti village via Sutol village and Nand Prayag, a service center.

The number of Khadu increases to four from Faldiya village. The Nesan of Latu Devta leads the procession up to Homekund from Wan village. After the procession goes to the highland from the last human settlements, Shrimad Bhagwat Katha starts in the Nanda Devi temple of Nauti village. People worship Goddess Nanda for the well-being and successful return of the devotees. A place called Rinkridhar is located on the way to Gairoli Patal. It is located in high slope gradient. It is believed that Shri Nanda Devi killed a demon in Rinkridhar. Dholidhar is located between Gairoli Patal and Bedini Bugyal. The entire area is covered by dense coniferous forest. Alpine meadows start at the ends of the tree line. Bedini Kund, a pious highland water body, is situated in Bedini Bugyal. The devotees take a holy dip in the Kund for self-renunciation. Tarpan (a ritual performed for worshiping forefathers) is also performed in the Kund. The importance of Bedini Kund can be noted from the fact that the Vedas are believed to have been written here. The Kund remains covered by

Fig. 4.16 Mount Trishuli, below which the Wan village, the last settlement on the Nanda Path, is located. Photo By author

snow during the winter and in the summer, numerous medicinal plants and flowers blossom surrounding it and in the meadows. The Nanda Ghunti and Trishuli, snow-clad peaks, can be seen from the Bedini Bugyal.

Patran Chaunya, also known as Niralidhar, is situated at 4000 m. It is 9 km from Bedini Bugyal. The devotees stay here overnight. An old story is associated with this place. Once, the King of Kannauj, Jaldhawal broke the tradition of the cultural procession of Goddess Nanda Devi. He allowed girls to dance in Patran Chaunya while he was participating in the procession. Shri Nanda Devi was angry with the King and the girls. She cursed the dancing girls to be converted into stone. After Patran Chaunya, another place, Shila Samundra (ocean of stones) lies 8 km from Bedini Bugyal. Shila Samundra is surrounded by mysterious rocks, spread over 2 km^2. The temples of Kaluwa Vinayak and Dhyo Singh, folk deities are situated in Shila Samundra. Roop Kund, Jyuraglidhar, Trishul, and Nanda Ghunti peaks can be seen from Shila Samundra. Bhaguwabasa, a picturesque alpine landscape, is situated 2 km from Shila Samundra. After a kilometer from Bhaguwabasa, Jyuraglidhar (5500 m) is situated, where pilgrims pay tax to Latu Devta. A highland mysterious glacial-fed lake named Roop Kund is situated at an altitude of about 5000 m, on the lap of Mt. Trishul, four km from Bhaguwabasa. The surrounding area of the lake is known for alpine flowers. The human skeleton of the fourteenth century AD is found in the lake. Roop Kund was discovered in 1942 by a forest ranger. The British official of India sent him to bring Alpine flowers. The forest ranger saw human skeletons for the first time in Roop Kund. Later on, Roop Kund was visited by many scientists in several times. The other halt of procession is in Chandniyaghat. The

Fig. 4.17 Elephant citrus is grown abundantly in the Pindar River basin on the Nanda Path. Photo By author

place is believed to be the meeting place of Lord Shiva and Goddess Shakti. Homekund is the last destination of the Nanda Path. The Dolies, Nesans, and Chhantolis along with Khadu are left in Homekund after the last puja of Nanda Devi. The Jagars are sung and Dev Dangers dance.

The Nanda Devi cultural procession depicts the rich cultural diversity of the Central Himalaya. People from all castes, creeds, and sects participate in the procession. Diversity in foods and beverages including fruits (Fig. 4.17) and vegetables can be noticed during the procession. At each halt, villagers offer food dishes to the folk deities and the devotees. The Nanda Path makes a picturesque landscape—river valleys, mid-altitudes, the highlands, pasturelands, and the snow-clad mountain peaks. The dense forests with rich biodiversity, crop fields, cascades, waterfalls, gorges, V and U-shaped valleys, wildlife, and avifauna further enhance the beauty of the landscape. People of all age groups and all castes and creeds participate in the procession.

4.3.4 Haridwar to Kotdwar, Pauri, and Srinagar

This route is renowned for the famous Shakti Temples—Jwalpa Devi and Dhari Devi. It is also known for the ancient sage Kanva whose temple lies close to Kotdwar city. It is known as Kanvashram, which is situated on the bank of Malini River. The journey starts from Haridwar or Hardwar—gateway to Badrinath (abode of Lord

Vishnu and Lord Shiva), respectively. Haridwar, situated on the bank of the Ganga River, the pious river of India, is district headquarters. It is also known as Moksha Dham as it is the biggest cremation center in India. People prefer cremation of their nearest relatives on the bank of the Ganga at Moksha Dham, Haridwar. Two Shakti Peeths—Mansa Devi and Chandi Devi are located on the hilltops, both sides of Haridwar.

The route goes through the dense forest of Rajaji National Park up to Nazimabad, a city of Uttar Pradesh State. Rajaji National Park has a rich biodiversity with dominance of deciduous monsoon forests and elephants. Furthermore, after Najibabad, dense forest of Rajaji National Park continues spanning up to Kotdwar (454 m), which is also known as Kotdwar Bhabar, a flat land. There are two rivers flow in Kotdwar—Malini and Khoh, provide ware to the city and irrigate agricultural land. The famous Siddha Bali Temple is situated on the left bank of the Khoh River, which remains open throughout the year and where devotees from the nearby villages visit whole year. Kotdwar is also known for the famous temple of Sage Kanwa (Fig. 4.18) Kotdwar serves the villages of its surrounding areas, a biggest market of the region. It has all types of educational institutions from primary to secondary and higher education. The road goes along the Khoh River through hilly terrain up to Dugadda (932 m), a service center, which it further bifurcates—one goes to Lansdowne (1780 m) and another to Gumkhal (1520 m). Lansdown is a famous hill town of Garhwal and known for the Army camp mainly for Garhwal Rifle. It is a tourist destination with pleasant climate during the summer and very cold during the winter. Gumkhal, a small service center, is located in the ridge with steep slope, surrounded by dense pine forests. The route between Dugadda to Gumkhal is

Fig. 4.18 Sage Kanva Ashram in Kotdwar, Pauri district

Fig. 4.19 (**a**) Pauri city and its spectacular landscape. (**b**) The Greater Himalaya is seen from Pauri

beautiful, close to which is Jaharikhal College. Pine forests make spectacular landscape in entire route. The road continues goes through the steep hills with ups and down. It is highly prone to landslide and road accident. After a long distance downslope, a city falls on the way is known as Satpuli (657 m). The city is located on the banks of the Nayar River. The settlement was washed away on September 14, 1951 due to massive flood in the Nayar River.

Furthermore, the road goes continues upwards and downwards. On the way, a famous Shakti Peeth Jwalpa Devi Temple is situated on the right bank of a perennial stream and on the road side. The temple remains open throughout the year and the pilgrims from different part of India in general and from the Garhwal Himalaya in particular visit the Shakti Peeth. Jwalpa Devi is an incarnation of Maha Kali. Navratris—Chaitra and Ashvin are celebrated and pilgrims worship folk deities. From Jwalpa Devi Temple, road goes through upslope along with dense pine forests and terraced agricultural fields and reached to Buwakhal (1550 m), a small service center. A road bifurcates to Khirsu, a hill town and famous tourism destination. Another road goes to Pauri city (1814 m), a district headquarters. The city landscape is panoramic and the climate is pleasant for the summer season, whereas the winters are cold. It is an educational, administrative, and business city (Fig. 4.19).

After 30 km from Pauri city, the Srinagar city (Fig. 4.20) is located on the stretch valley and the banks of the Alaknanda River. Just after another 30 km from Srinagar is located the Dhari Devi Shakti Peeth. This is the end of the journey.

The entire route makes rich natural and cultural diversity. Landscape is spectacular—the river valleys, middle altitudes, and the high Himalaya. Forest landscape varies from monsoon deciduous to pine, mixed-oak and coniferous forests. Fauna and avifauna diversity is also rich. The most importantly, agro-diversity is rich. Crops grown are sugarcane in the plain regions, paddy and wheat in the river valleys, mixed-cropping domination by food grains, pulses, and oilseeds in the middle and highlands. Rich diversity is noted in vegetables and fruits, vary according to slope gradient. The region is also rich in performing culture, customs, rituals, and celebrating fairs and festivals.

Fig. 4.20 Srinagar city is situated on the left bank of the Alaknanda River, making a panoramic view. Photo By author

4.4 Conclusions

In this chapter, the legend of Kedarkhand is divided into four parts, covering it as a whole, and leading to the Himalaya through different trails. The first part described is Haridwar to Badrinath and Kedarnath. It describes all natural and cultural aspects of the route. The second route is traversed from Rishikesh to Gangotri and Yamunotri through Uttarkashi. This route is rich in culture, famous for the two highland pilgrimages—Gangotri and Yamunotri. The third route leads to Homekund, Sutol, and Nand Prayag from Karnprayag. This route make a complete circuit. The entire route is called the Nanda Path, a folk deity, which goes to Kailash, the abode of Lord Shiva, every year. The profession is followed by the priests, devotees, and tourists from all over India. The last route is traversed from Haridwar to Pauri and Srinagar cities. Two famous Shakti Peeths are located in this route. Furthermore, the valley pilgrimage Haridwar is the part of this cultural route. The culture, customs, rituals, languages, clothes, foods, and beverages vary in all four part of the Kedarkhand. All cultural and agricultural practices are traditional.

References

Anshu SK, Pandey VK (2024) Impact of climate change on agricultural activities in garhwal himalayas: a comprehensive analysis on shifts, changes and adaptation strategies. J Geogr Environ Earth Sci Int 28(2):1–10. https://doi.org/10.9734/jgeesi/2024/v28i2746

Bharati A (1965) The tantric tradition. Rider and Company, London

Brooks DR (1990) The secret of the three cities: an introduction to Hindu Shakta Tantrism. University of Chicago Press, Chicago

Chakrabarty S (2015) Sacred spaces in Indian religions. Routledge, London

Chand S, Mishra L (2024) Pilgrimage tourism in Uttarakhand: a critical review with reference to the Char Dham Yatra. Int J Res Publ Seminar 15(1). https://doi.org/10.36676/jrps.v15.i1.1647

Chanyal S, Sharma N, Srivastava R K, Chhimwal M (2019) Necessity of dairy farming in Gairsain area of Garhwal Himalayas. Int J Livestock Res 9(10):185–191

Dobhal DP, Gupta AK, Mehta M, Khandelwal DD (2013) Kedarnath disaster: facts and plausible causes. Curr Sci 105(2):171–174. https://www.jstor.org/stable/24089103

Garuda Purana. (1908) Hindu Scriptures Series, translated by Manmatha Nath Dutt, Elysium Press.

Gopal R, Bali RS, Prajapati DR, Rathod DU (2025) Traditional agroforestry systems in Garhwal Himalaya. J Plant Dev Sci 9:793–797

Jain SK, Goswami A, Saraf AK (2017) Flood hazards in the Himalayan region: a case study from Uttarakhand. Int J Disaster Risk Reduct 21:124–133. https://doi.org/10.1016/j.ijdrr.2016.11.011

Kakkar G, Mishra A (2025) Analyzing tourism infrastructure in Indian Himalayan Region states. Int Adv Res J Sci Eng Technol. https://doi.org/10.17148/IARJSET.2025.12411

Kinsley D (1997) Tantric visions of the divine feminine: the ten mahavidyas. University of California Press, Oakland

Lingwal SN, Uniyal MC, Lingwal NS (2022) Role of Garhwal cuisine in promoting tourism in the region. Int J Spl Educ 37(3):1625

McDaniel J (2004) Offering flowers, feeding skulls: popular goddess worship in West Bengal. Oxford University Press, Oxford

Nagar B, Kumar M, Pandey R, Rawat S (2025) Fuelwood and fodder consumption from agroforestry at different altitudinal zones of Garhwal Himalaya. J Plant Dev Sci 12:659–668

Rana P, Pant PD, Pandey A, Bhatt CM (2022) Cloudburst-induced flash floods and landslides in Uttarakhand Himalaya: a geoenvironmental appraisal. Geoenviron Disasters 9(1):1–17. https://doi.org/10.1186/s40677-022-00208-3

Sati VP (2004) Horticultural development in hills: a case for the Alaknanda Basin. Mittal Publications, New Delhi

Sati VP (2012) Agricultural diversification in the Garhwal Himalaya: a spatio-temporal analysis. Sustain Agric Res 1(1):77–85. https://doi.org/10.5539/sar.v1n1p77

Sati VP (2021) The ganges: cultural, economic, and environmental significance. Springer, Cham

Sati VP (2024) Farming systems and sustainable agriculture in the Himalaya. Springer, Cham

Shastri JL (1983) The shiva purana: translations and commentaries. Motilal Banarsidass Publishers, New Delhi

Shrestha AB, Eriksson M, Mool P, Ghimire P, Mishra B, Khanal NR (2010) Glacial lake outburst flood risk in the Himalayas: mitigation strategies and preliminary results. Ann Glaciol 50(53):53–62. https://doi.org/10.3189/172756410790595804

Singh D, Semwal M, Semwal MM (2024) Analysis of Uttarakhand tourism policy 2023: a sustainable tourism perspective. J Mountain Res 19(2):67–75

Sivananda S (2003) The path of bhakti yoga. Divine Life Society, Rishikesh

Skanda Purana (1968) Kedarkhand. Translations and Commentaries, Motilal Banarsidass

Vayu Purana (1980) Critical edition by Anand Swarup Gupta. All-India Kashiraj Trust, Varanasi

Verma K, Rawat A, Dhodi RK, Dhodi R (2024) Community participation in sustainable tourism development: An application of MOA model in the Garhwal Himalayas of Uttarakhand, India. Int J Tour Policy 1(1). https://doi.org/10.1504/IJTP.2024.10064169

References

Chapter 5
Manaskhand: The Climax of Nature and Culture

Contents

Abstract Manaskhand is the ancient name for the Kumaon Himalaya, which is further divided into two parts: western Kumaon and eastern (Kali) Kumaon. Manaskhand represents the culmination of both natural beauty and cultural heritage. The region is home to numerous natural locales of significant economic importance. Additionally, Kumaon boasts a rich cultural landscape, featuring renowned temples such as the Sun Temple, Jageshwar Jyotirlinga, Kasar Devi, Bageshwar, and Adi Kailash. While western Kumaon is known for the Gomati and Saryu river valleys, eastern Kumaon is distinguished by the Johar, Darma, and Vyas valleys. Notable natural sites in the region include Kausani, Nainital, and Munsiyari. This chapter presents a vivid description of the culture, economy, and environment of the area.

Keywords Jageshwar · Munsiyari · Panchachuli · Adi Kailash · Kali Kumaon · Manaskhand

V. P. Sati, *Exploring the Mystical Himalaya*, Springer Geography,
https://doi.org/10.1007/978-3-032-14275-7_5

5.1 Introduction

Manaskhand is an ancient region with deep cultural, religious, and historical significance, primarily located in the Himalayan belt of Uttarakhand, India (Fig. 5.1). It is mentioned in various Hindu scriptures and texts, particularly the "Skanda Purana," one of the largest Maha Puranas, where it describes the divine geography of the region, linking it to various sacred sites and pilgrimages (Bhattacharyya 1922). Manaskhand is revered for its connection to Lord Shiva and the presence of holy shrines, temples, and rivers (Doniger 1991). The region is known for its sacred temples such as Jageshwar, Katarmal, and Patal Bhuvaneshwar, where devotees gather to offer prayers and perform rituals (Eliade 1987). The religious importance of the area is not limited to these temples, as the entire landscape is considered blessed by the presence of deities and ancient sages.

In addition to its spiritual aspects, Manaskhand is renowned for its natural beauty, with dense forests, towering peaks, and pristine rivers forming part of the Himalayan ecosystem. The rich biodiversity, including wildlife and endemic flora, adds to the significance of the region, making it not only a spiritual destination but also a haven for nature enthusiasts and trekkers. The cultural heritage of Manaskhand is preserved through local traditions, festivals, and practices that have been passed down for generations. It is a region where mythology and history blend with the natural environment, offering insights into the deep-rooted connection between people and nature in the Himalayan context (Ramesh 1996).

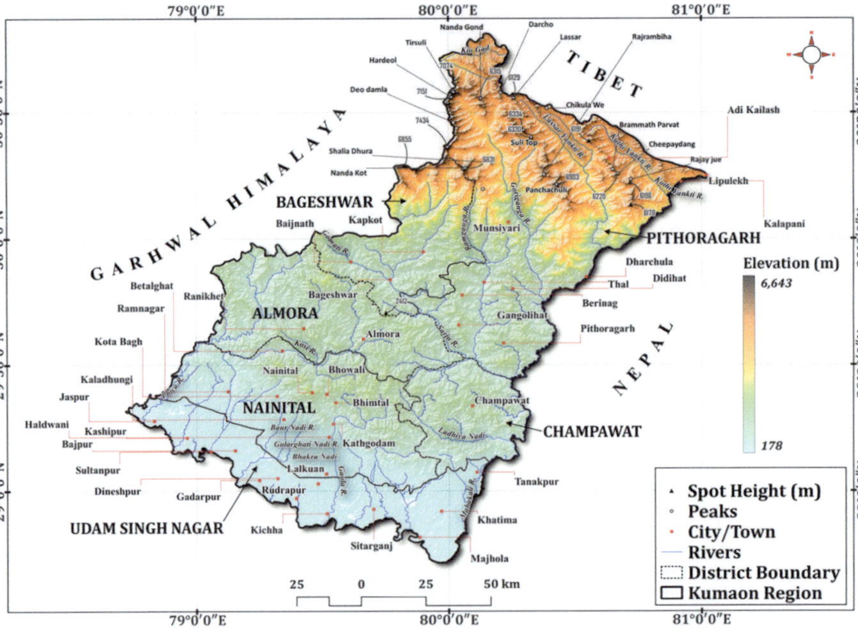

Fig. 5.1 Location map of Manaskhand (Kumaon Himalaya)

The history of the Manaskhand is rich and ancient. It is referred to in the ancient treatises of Ramayana and Mahabharata. Adi Shankaracharya visited Bageshwar and Jageshwar in the beginning of the ninth century, and he constructed and renovated the temples of Bageshwar and Jageshwar (Sati 2022). He is believed to be a pioneer of trekking in the Himalayas because he trekked these places. Two dynasties—Katyuri and Chand—ruled the Manaskhand during different periods. The establishment and construction of most of the temples in the Manaskhand were done by Katyuri rulers. They were reformers and conservators of culture. Recently, cultural integration is being carried out through cultural processions. Adi Kailash is now visited by a number of pilgrims from India and abroad (Sood 2003). The route to Kailash Mansarovar goes through the Kumaon Himalaya, and as a result, all cultural processions to Kailash Mansarovar move through Manaskhand (Vogel 2011).

Manaskhand is the ancient name of the present Kumaon Himalaya, extending from Gwaldam, located in the middle western part of the Kumaon Himalaya, to Kali Kumaon along the Kali River, which divides the Kumaon Himalaya from Nepal. At present, there are six administrative districts, including Bageshwar, Almora, Nainital, Udham Singh Nagar (USN), Champawat, and Pithoragarh from west to east. The northern part, mainly three districts—Pithoragarh, Almora, and Bageshwar—is snow-capped. Many rivers originate and flow, which are part of the Kali River system. Among them, Saryu, Gomati, Ramganga (West and East), Dhauli Ganga (East), Gori Ganga, and the Kali River are prominent (Sati 2021). These rivers are glacier-fed, perennial, with high volume and velocity. The entire region is rich in cultural and natural diversity. It is a land of Gods and Goddesses, Veers, and Ghosts, which are worshiped by local people. Many Shiva Temples, Shakti Temples, and Vishnu Temples are located in the Kumaon Himalaya. The famous temples include the Sun Temple at Katarmal, the Kasar Devi Temple and the Golu Temple near Almora city, the famous temples of Jageshwar (30 km from Almora), Chhota Kailash in Pithoragarh district, Purnagiri Temple in Tanakpur, and Dunagiri Temple in Dwarhat. These are the famous pilgrimages of the Kumaon Himalaya. This region is also rich in numerous perennial highland and lowland lakes. Nainital district has seven natural lakes.

Forest resources are abundant, and the region has rich biodiversity. Since altitude varies from below 300 m to above 7000 m, forest types also vary from subtropical monsoon forests to pine, mixed-oak, coniferous, and alpine grasslands (Sati 2023). These forests have high economic value. Corbett National Park, Binsar Wildlife Sanctuary, and Sona Nadi Wildlife National Park are among the prominent national parks and wildlife sanctuaries in the Kumaon Himalaya, which provide bases for eco-tourism. Agrobiodiversity is rich, with more than 20 crop races grown in this region. The agricultural lands are terraced, and agriculture is practiced in a traditional mode with subsistence production. Varieties of fruits, vegetables, food grains, and spices are grown using organic manure, and therefore the nutritional value of these crops is substantial (Sati 2024).

5.2 Agriculture, Tourism, and Sustainable Livelihood

Agriculture is the main occupation and a major source of livelihood in the Manaskhand region (Sati 2025). Here, about 70% of people are engaged in cultivating traditional cereal crops. Agrobiodiversity is high, and many crop races are grown (Rawal and Dhar 2021). Recently, the cropping pattern has been changing, which has had a significant impact on food security in the Kumaon Himalaya (Joshi et al. 2022). This shift has also led to a decline in crop production and productivity. Climate change is one of the major drivers of this decline. To cope with climate change and food insecurity, climate-smart agriculture is needed. Furthermore, policy support is essential to enhance crop production and productivity through climate-smart agriculture (Pant and Pande 2023).

The livelihood of the people is primarily dependent on biomass-based agriculture, livestock farming, and the collection of forest products. However, these economic activities are insufficient to meet even the basic food requirements of the local population. Therefore, integrating traditional and modern agricultural systems is necessary to sustain livelihoods (Negi et al. 2020). One of the key policy measures is the diversification of rural livelihoods—from traditional cultivation to the cultivation of cash crops, dairy farming, sustainable use of forest resources, and the adoption of modern innovations in agriculture (Mehta and Tiwari 2021). The Kumaon Himalaya is also experiencing migration and changing rural livelihoods; consequently, land abandonment has become common (Sharma and Joshi 2024).

Tourism development has significant potential for economic growth and sustainable livelihoods. The Kumaon Himalaya is endowed with high-quality tourist destinations, especially for nature and adventure tourism. The Johar Valley, Darma Valley, Vyas Valley, and Koshi Valley are among the major, though remotely located, trekking routes. The region also hosts several national parks and wildlife sanctuaries, which can be developed for eco-tourism (Nautiyal and Rao 2023). Prominent pilgrimage sites in the Kumaon Himalaya include Adi Kailash, Jageshwar, and Bageshwar, which attract thousands of pilgrims every year. Many trekking routes lack adequate accommodation facilities; hence, homestay tourism can be promoted as a sustainable alternative (Tiwari and Bhatt 2021). Furthermore, community participation in tourism development at key destinations is essential for sustaining tourism (Bisht and Dangwal 2022).

5.3 Natural and Cultural Significance

The Manaskhand is divided into two natural and cultural realms—the western region and the eastern region. The western region includes the area from Gwaldam, a town in Chamoli district located in the transitional zones of the Garhwal and Kumaon Himalayas, to Bageshwar (Bageshwar district), Jageshwar (Almora district), and Nainital (Nainital district), including many other cultural and natural locales. The second region includes mainly Champawat and Pithoragarh districts with major natural and cultural destinations. Three valleys are famous: the Johar Valley, located in the Gori Ganga basin, where the famous Uttarakhand city of Munsiyari is located; the Darma Valley, located in the Dhauli Ganga East basin; and the Vyas Valley, located in the extreme northeast part of the Pithoragarh district, where the famous Panchachuli mountain peak is located. The Vyas Valley is renowned for Chhota Kailash and Om Parvat. The description of both regions is based on my extensive visits and observations.

5.4 The Western Kumaon

In the beginning of 2023, I visited the magnificent Kumaon hills. The journey started from Dehradun to Nainital via Haridwar, Najibabad, Dhampur, Kashipur, Bajpur, Kaldhungi, Corbett Fall, Ghatgarh, Mangoli, Khurpa Tal, and Sariya Tal. It was about a 200-km journey, covered in a day. From Dehradun to Kaldhungi, the road goes through plain areas covering Uttarakhand and Uttar Pradesh states. The world-famous Jim Corbett National Park is located about 40 km from Kaldhungi, before reaching Nainital. The hilly terrain begins from Kaldhungi, and along the road, dense pine and oak forests are found up to Nainital.

It was winter season, and since Nainital is located at an altitude of 2000 m, it was very cold. We got accommodation at the Kumaon University guest house named Panchachuli. Panchachuli is the highest snow-clad mountain peak (Himalaya) of the Kumaon Hills. Nainital is known for its beautiful perennial lake, called Naini Lake (Fig. 5.2), and the Mall Road, surrounded by dense oak and Deodar forests. The famous temple of Naini Devi is situated on the bank of Naini Lake in its upper part, called Mallital. The Mall Road remains crowded from morning to late evening due to the presence of tourists throughout the year. A number of hotels, dhabas, and shops serve the tourists. The food quality in Nainital was excellent with lots of Pahadi flavor. Two days are needed to visit Nainital‘s surrounding areas.

Nainital and its surrounding areas have many famous tourist destinations, such as China Peak (accessible by a 5-km taxi ride and 3-km trek) and seven perennial lakes. The Himalayan ranges are visible from China Peak, making the entire landscape beautiful. Outside of Nainital town, there are many natural lakes: Khurpa Tal, Sariya Tal, Sat Tal (comprising Nal Damyanti, Hanuman Tal, Garur Tal, Ram Tal, Sita Tal, Lakshman Tal, and Bharat Tal), Bhimtal Tal, and Naukuchiya Tal. Boating and adventure tourism activities are available at Sat Tal. All lakes have spectacular landscapes. Within Nainital town, there are many quality tourist destinations. The Mall Road and the surroundings of Naini Lake are worth visiting.

The journey goes from Nainital to Jageshwar via Bhawali, Mukteshwar, and Almora. There are two routes to Jageshwar from Nainital: one goes directly to Almora, and the second goes through Mukteshwar. The natural and cultural landscapes of both routes are spectacular. Situated on a mountain peak at 2400 m altitude, Mukteshwar (Fig. 5.3) is located approximately 48 km from Nainital. Since

Fig. 5.2 Naini Lake at Nainital town (1900 m) in the Kumaon Himalaya

Fig. 5.3 Mukteshwar temple (2400 m), located in the Almora district between Nainital and Almora cities

Mukteshwar is located off-route, the road condition was not good, although the landscape along the way was spectacular. It took more than 2 h to reach Mukteshwar.

The temple of Mukteshwar is devoted to Lord Shiva and is surrounded by dense deodar trees. The mighty Himalayan ranges can be seen from here. Mukteshwar is a place where one can get rid of the cycle of birth and death. It is believed to be a gateway to liberation or heaven. At the foot of Mukteshwar peak, small shopkeepers sell citrus fruits, mainly oranges and lemons, and juices. Along with fruits, they also serve tasty Pahadi foods. Mukteshwar was indeed an enjoyable journey. Between Mukteshwar and Almora, a famous wildlife sanctuary——Binsar—is located, which is known for its faunal and floral diversity and beautiful natural and forest landscapes.

It was about a 50-km journey from Mukteshwar to Almora via the dense forests of Binsar Wildlife Sanctuary. The road condition was poor, so it took about 2 h to reach Almora. From Almora, the Kasar Devi temple is located about 7 km away at an altitude of about 1900 m. The temple of Kasar Devi is beautiful, and worshiping the deity provides peace of mind and tranquility. The place is believed to have electromagnetic waves under the influence of the deity. Swami Vivekananda visited the temple and performed meditation in a cave close to it. After meditating on the temple premises, Swami Ji began serving the downtrodden people. There was deep peace and tranquility while meditating in the premises of the Lord Shiva temple, situated on the hilltop about 50 m from Kasar Devi temple. The temples are surrounded by pine forests, enhancing the cosmic beauty.

The sun was setting slowly. Jageshwar Dham was about 35 km from the Kasar Devi temple. After traveling a few kilometers, we encountered the Golu Devta temple on the way to Jageshwar. Pilgrims visiting Jageshwar halt here to worship Golu Devta for his blessings. Thousands of brass bells—small and big—hang in the temple premises, some weighing more than 100 kg. Along with bells, there were thousands of posted letters. While discussing the bells and letters with the temple priest, he explained that devotees of Golu Devta offered the bells after their wishes were fulfilled. Many devotees write letters to Golu Devta seeking solutions to their problems. It was a miraculous place where one feels peace of mind and tranquility. There are many shops serving tea, snacks, and traditional foods to pilgrims. From here, Jageshwar Dham is 20 km downslope. The road descends through changing forest patterns—pine forests at higher altitudes and deodar forests in the lower elevations and around Jageshwar Dham (Fig. 5.4). Interestingly, this was contrary to the usual pattern, as deodar forests typically grow at higher altitudes (above 2000 m) while pine forests are found below them (800 m to 1800 m).

It was late evening. The surrounding area of Jageshwar Dham was deeply dark with only a few street lights on the roads. Jageshwar is a small town situated at 1400 m on the bank of the Jataganga and surrounded by dense deodar forest. It is famous for its group of 124 temples dedicated to Lord Shiva and other folk deities. Jageshwar is one of the Jyotirlingas established by Adi Shankaracharya in the ninth century. There are many hotels and restaurants with sufficient rooms for accommodation and food and beverages for pilgrims. The weather was very cold, with temperatures dropping below 0 °C (Bisht 2016). Early in the morning, a sheet of frost covered everything, making the roads slippery. The group of 124 temples is situated within the core area of Jageshwar town. The Puja of Shiva Lingam and Shakti (Parvati) began at 7:00 am. Pilgrims came to the temple premises in groups to worship Lord Shiva. On the other side of the Jageshwar temple premises stood a temple

(a) (b)

Fig. 5.4 (**a**) Jageshwar Temple. (**b**) A devotee welcoming pilgrims with playing Dhol and Damau, traditional musical instruments at the entrance gate of Jageshwar Temple

of Lord Kuber (Lord of wealth). A priest at the Kuber temple was performing the morning chant. In the main temple premises, some priests continued to burn wood while others performed Puja. The smoke from the burning wood covered the temple premises. Monkeys roamed here and there, snatching food items from the devotees. The entire atmosphere was filled with peace and tranquility. The sound of Vedic hymns pronounced by the priests was clearly audible. The sacred water of Jataganga was frozen and still. Outside the temple premises, shopkeepers sat circling fire pots, selling goods to devotees for offerings to Lord Shiva.

The same route leads back to Almora. However, before Almora, a bypass road leads to Kosi, which is about 30 km away. Kosi is a town and a small service center situated on the bank of the Kosi River. About 3 km from Kosi, the famous Surya Temples are situated on a sun-facing slope above Katarmal village. The temples are magnificent and awe-inspiring. From Kosi, a road bifurcates—one route goes through Gwaldam, while the other passes through Ranikhet. Both roads meet at Simli, a service center situated on the bank of the Pindar River. About 30 km away lies Kausani, a popular natural tourism destination. The river valley between Kosi and Kausani is known as the Someshwar Valley, a region rich in cultural and natural heritage. "Someshwar"is another name for Lord Shiva. Kausani, the hometown of the renowned poet Sumitranandan Pant, has been visited by many Indian and foreign scholars over the years. A long chain of the mighty Himalayan mountains can be seen from here. About 10 km below Kausani are the towns of Garur and Baijnath, located in the Garur Valley. The Gomati River flows through this valley. A group of 24 temples dedicated to Shiva and Shakti are situated on the right bank of the Gomati River in Baijnath (Fig. 5.5).

These spectacular and ancient temples, known as the Baijnath Temples, are believed to have been constructed by the Katyuri rulers. Tourists and pilgrims visit these temples throughout the year. From Baijnath, one road leads to Gwaldam, a border town between Kumaon and Garhwal, while another road leads to Bageshwar, located about 20 km along the Gomati River. Bageshwar is a culturally significant town in the Kumaon Himalayas, situated at the confluence of the Gomati and Saryu rivers. The famous Baghnath Temple, dedicated to Lord Shiva, is located here. Bageshwar is also known as Triveni, as it is believed that the Bhagirathi River once flowed here, though it has now disappeared. The town hosts many fairs and festivals, the longest being the Makar Sankranti fair, which continues for 20 days. The Baghnath Temple was constructed during the Katyuri rule in the Kumaon Himalayas. Adi Shankaracharya is said to have visited this place and renovated the temple. Gwaldam (2000 m), another important place on this route, is known as a sleepy town. Close to the mighty Himalayas, particularly Mount Trishul (Trishuli), Gwaldam is a lush green and picturesque town. It has a unique blend of Garhwali and Kumaoni cultures.

Another route from Kosi leads to Ranikhet, which is about 45 km away, a journey of approximately one and a half hours. The road winds through several curves, flanked by dense pine forests on both sides. The landscape along the way is spectacular. Ranikhet is a beautiful hill town and an army cantonment. The climate is very pleasant during the summer, while winters are severely cold, with occasional snowfall. From Ranikhet, the road descends to Dwarahat, where it further bifurcates—one route leads to Ramnagar, while the other goes to Karnprayag. The Dwarahat-Karnprayag-Dehradun route is a long journey through mountain peaks and river valleys, passing through Panduwakhal, Mahalchauri, Gairsain, Diwali Khal, Adi Badri, and Simli. Along the way, there are many hotels and restaurants available for accommodation and food. Panduwakhal serves as a service center and marks the political boundary between Garhwal and Kumaon. Adi Badri, one of the Panch Badris, is located on the banks of the Ata Garh River. It features a group of

Fig. 5.5 Famous Baijnath Temple, located on the bank of the Gomati River on the way to Bageshwar town

temples dedicated to Lord Vishnu, believed to have been constructed by Adi Shankaracharya.

The road continues through Karnprayag, Gaucher, Ratura, Rudraprayag, Srinagar, Devprayag, and Rishikesh. Karnprayag, where the Pindar River meets the Alaknanda River, Rudraprayag, where the Mandakini River joins the Alaknanda, and Devprayag, where the Bhagirathi and Alaknanda Rivers merge to form the Ganga, are three important cultural and religious sites. Before reaching Srinagar, a culturally significant city of Garhwal, the Dhari Devi Temple is located. Following the construction of a dam reservoir on the Alaknanda River at Koteshwar near Srinagar Garhwal, the Dhari Devi Temple was reconstructed in the middle of the reservoir. Rishikesh, known as the "Yoga Capital of the World" and a sacred pilgrimage valley, is situated on the banks of the Ganga River. Finally, we reached Dehradun, passing through a beautiful stretch of dense monsoon deciduous forests in Rajaji National Park.

On this journey, we visited two national parks—Jim Corbett and Rajaji National Park—along with seven cultural sites: Naini Devi Shaktipeeth, Jageshwar Jyotirlinga, Kasar Devi Temple, Mukteshwar, Surya Temple, Adi Badri, and Dhari Devi Shaktipeeth. Throughout the way, we enjoyed breathtaking views of the snow-clad Himalayan peaks. In summary, this journey was a perfect blend of natural beauty, wilderness, and cultural richness, offering a remarkable experience at the intersection of nature and heritage.

5.5 The Eastern (Kali) Kumaon

The eastern part of the Kumaon Himalaya is also known as Kali Kumaon, largely influenced by the Kali River, which divides the two macro-cultural realms of eastern Kumaon and western Nepal. The Kali River, also known as Choka and Sharda, serves as the natural boundary between Kumaon and Nepal. This region is renowned for Adi Kailash (Chhota Kailash), Om Parvat, and the famous Jauhar, Darma, and Vyas Valleys. The journey along this route begins in Bageshwar town and passes through Kapkot, Munsiyari, Joljivi, Dharchula, Tawa Ghat, Lohaghat, Champawat, Pithoragarh, and Tanakpur.

It was the winter of 2024. My university had closed for a month, so I returned to my hometown, Dehradun. For many years, I had wanted to visit the Kali Kumaon region, home to Adi Kailash, but circumstances had never allowed me to do so. In fact, while writing my book on the Himalaya, I had left out the Kali Kumaon region. This time, I was determined to finally visit Adi Kailash. On a Wednesday, I set out from Dehradun for the journey. The night before, heavy snowfall had blanketed the high-altitude areas, while the lower valleys had received rain. My wife advised me to reconsider the trip, but my mind was made up. That morning, at 7 am, I departed from my home, "Nilaya Hills." Light drizzle accompanied me throughout the journey, and in the upper regions, a thick blanket of snow was visible. My first stop was Gwaldam, approximately 280 km from Dehradun. The road from Dehradun to

Rishikesh was both scenic and exciting, passing through the dense forests of Rajaji National Park. From Rishikesh, I traveled along the banks of the Ganga River via Vyasi and reached Devprayag. On the way, I stopped at a place called Teen Dhara for breakfast. Continuing my journey, I followed the Alaknanda River, passing through Devprayag, Srinagar, Rudraprayag, and Karnprayag. Upon reaching Karnprayag, I still had about 80 kilometers to cover. I proceeded at a steady pace and finally arrived in Gwaldam at 4 pm. Since the Garhwal Mandal Vikas Nigam tourist lodge was available at a reasonable rate, I decided to stay overnight. I booked a deluxe room for INR 1200 and settled in for the night.

Traveling 280 km by road was an enjoyable experience. The journey unfolded through breathtaking river valleys, towering mountains, dense forests, terraced fields, granaries, and a mix of rural and urban settlements. Along the way, the local food and beverages were exceptionally delicious. Gwaldam, a picturesque village and service center, has retained its charm since the British era. Surrounded by dense deodar forests on two sides and tea plantations and rural settlements on the other, it sits at an altitude of approximately 2000 m. In winter, Gwaldam experiences snowfall, transforming it into a serene, snow-covered retreat. The residents primarily engage in trade for their livelihood, and during summer, the village becomes a popular tourist destination in Uttarakhand. A morning or evening walk in Gwaldam is a refreshing experience, offering both tranquility and rejuvenation. From here, the majestic Himalayas and the Trishuli peak are visible, while the cold winds blowing down from the mountains create a shivering sensation. To withstand the chill, warm clothing is essential. The region is abundant in fresh, organic fruits and vegetables. Malta oranges, large lemons, and dairy products are widely available. Traditional meals, such as chapatis, lentils, and potato-cauliflower curry, are both plentiful and incredibly delicious. Gwaldam is often called the "Sleeping Village" due to its slow-paced life. Shops close by 6 pm, with only one or two remaining open until 8 pm to serve dinner. The culture here is a blend of Garhwali and Kumaoni influences, reflected in both the language and local traditions.

During winter, the morning temperature in Gwaldam often drops to zero, with the climate heavily influenced by the Himalayas, particularly Mount Trishul, which stands prominently in front of the village. The panoramic view of the Greater Himalayas from Gwaldam is breathtaking, attracting tourists who eagerly capture photographs from early morning until midday. The snow-clad peaks appear even more stunning after fresh snowfall. About 20 km from Gwaldam lies Baijnath, a town situated on the banks of the Gomati River. Baijnath is renowned for its 24 ancient temples dedicated to Shiva and Parvati, believed to have been built by the Katyuri kings. The town is nestled in a valley with fertile terraced agricultural fields (Pant 2010). From Baijnath, the road bifurcates—one route leads to Kausani, while the other heads to Bageshwar, located approximately 20 km away. Bageshwar is famous for the Baghnath Temple, dedicated to Lord Shiva, who is locally worshipped as Baghnath. The temple is situated at the confluence of the Gomati and Saryu rivers, at an altitude of 1100 m above sea level. The valley of the Gomati River between Baijnath and Bageshwar is also known as the Garur Valley, a region known for its scenic beauty and agricultural richness. From Bageshwar, a road follows the left bank of the Saryu River to Kapkot, which is about 20 km away. From Kapkot, one route continues toward Thal, while another leads to Munsiyari. This entire region is historically known as Danpur.

From Kapkot, the journey takes a steep ascent through dense cedar forests, eventually reaching Jhopada village (Fig. 5.6). Along the roadside, a small tea shop offers tea, coffee, and snacks, providing a much-needed respite for travelers. The road winds through breathtaking river valleys, middle-altitude zones, and towering mountain ranges that seem to touch the sky. In certain sections, the path narrows significantly, with deep gorges below, evoking a constant sense of awe and apprehension. Despite the challenges, the entire route is extraordinarily scenic—adorned with terraced fields, remote rural settlements, and vast cedar and deodar forests. The landscape feels as though it has been sculpted by divine hands, leaving travelers deeply moved by its sheer beauty.

The road begins its descent once again, passing through Khadlekh (2020 m), Sama, and Masuri Nagar (1180 m) before advancing along the eastern Ramganga River. Soon, it starts climbing to higher altitudes, traversing Devana, Sweeti (1180 m), Fully, Jhekha (1650 m), and several other important rural settlements along the roadside. One of the highlights of this stretch is Birthi Falls (Fig. 5.7), a breathtaking waterfall cascading from the Himalayas. With a height of over 200 m, this perennial waterfall is framed by steep rock cliffs on both sides, enhancing its dramatic beauty. Tourists frequently stop here to relax, take photos, and capture videos of the mesmerizing scenery. A 1-km trail leads up to the waterfall, offering visitors a closer experience of its grandeur. Numerous hotels and restaurants in the area provide food and accommodation for travelers. While some tourists stop here

Fig. 5.6 Jhopada village on the way to Munsiyari town in Pithoragarh District, Kali Kumaon

briefly before continuing their journey, others choose to stay overnight to fully appreciate the tranquil surroundings.

After Birthi Fall, a place called "Ratapani" (2500 m) is located, where a small army camp office lies on a steep slope of hard rock. At the highest point, Kala Muni Top (2750 m) is situated, where a temple dedicated to Goddess Kali stands. It was covered in snow. From the temple premises, a trail leads to Thamari Kund, a natural spot. It is a beautiful place where tourists rest, enjoy the scenery, take photographs, and make reels. About 5 km from here, there is another spot called Khaliya Dwar. Due to its high altitude, it remains covered in snow during the winter. Descending further is the town of Munsiyari (Fig. 5.8), which is often referred to as the Switzerland of India. From Dehradun to Munsiyari, there are seven river valleys—the Ganga, Alaknanda, Pindar, Gomati, Saryu, Ramganga (East), and Gori Ganga.

Right in front of Munsiyari lies the Panchachuli Peak, covered in thick snow. Along with it, the long range of the Himalayan mountains can also be seen. It is believed that the setting of Munsiyari resembles that of Zurich. The average altitude of Munsiyari is 2100 m, but the temperature drops below 0 °C in winter. There are many hotels and restaurants where traditional food and beverages are served to tourists. The staple food includes local rice, rajma, potato-based dishes, and Chaat. These items are nutritious, healthy, and tasty. The winter evenings in Munsiyari are very cold, as it receives heavy snowfall. The Panchachuli group of mountains further accentuates the severity of the cold.

There are many natural, cultural, and tourist destinations in both western and eastern Kumaon. Munsiyari town is one of them. Located in the Gori Ganga basin, the upper part of which is known as the Jahar Valley, Munsiyari is a small town

Fig. 5.7 Birthi fall (200 m height) before Munsiyari

Fig. 5.8 Munsiyari town, renowned as the Switzerland of Uttarakhand, located in the Johar valley of Kali Kumaon

serving both tourists and the people of the surrounding areas. The Gori Ganga (Fig. 5.9) originates from the Milam Glacier, contributing to the spectacular landscape. The cityscape of Munsiyari is gently sloped, facing the Panchachuli peaks and surrounded by trees. The upper reaches are covered with coniferous trees, while pine trees are found in the lower basin. The Johar Valley is also known for its tribal population, mainly the Shauka, who belong to the Indo-Tibetan human race (Keith 1917). The Shauka have two dwellings: during the winter, they seasonally migrate to lower-lying areas, and during the summer, they migrate to the highlands of the Johar Valley. They grow rajma and potatoes, and they rear goats, sheep, lambs, and horses. As nomads, they graze their animals in the alpine grasslands, known as Bugyals in Uttarakhand.

The trail descends from Munsiyari to the bank of the Gori Ganga. The road condition is poor, as the landscape is fragile and the road is narrow. Small landslides are common along the road, while the river valley offers panoramic views. Approximately 10 km from Munsiyari, one trail leads to the Milam Glacier, about 80 km away, while another trail heads to Joljivi, where the Gori Ganga joins the Kali River. Joljivi is approximately 70 km from Munsiyari. The entire valley is rich in biodiversity, both faunal and floral. Agrobiodiversity is also abundant, with the Gori Ganga basin growing many crop varieties, including food grains, pulses, fruits, and vegetables (Kashyap 2018; Rawat 2007). The Gori Ganga Valley is lush and green, adorned with countless waterfalls, gorges, and meanders. The terraced fields, traditional rural settlements, and the natural beauty of the valley add to its charm. Along the way, several rural settlements are located along the Gori Ganga, such as Dharkot

Fig. 5.9 The Gori Ganga catchment. The Gori Ganga originates from the Milam Glacier and joins the Kali Ganga in Joljivi

(1800 m), Bangapani (1050 m), Baram, and Garja, before reaching Joljivi (1100 m). Some spots along the route are breathtaking. In several places, the road passes through waterfalls, with water cascading around it.

We conducted a study of Bangapani, a beautiful village (Fig. 5.10), which is rich in agrobiodiversity. We interviewed a local resident, Bhupender Singh Parihar. The residents of this village are marginal farmers, and their livelihoods depend on traditional agriculture and livestock rearing. They grow many crop cultivars and races using organic manure. They also rear cows and buffaloes, selling milk in the local market to sustain their livelihoods (Gupta 2014; Sharma 2015). There are many other villages located along the Gori Ganga Valley. After a certain distance, the urban center of Joljivi lies, where the Gori River merges with the Kali River. This confluence is known as the meeting point of the Gori and Kali rivers. The Kali River was fierce and immense—truly embodying its name. Joljivi is also home to the Jwaleshwar Temple, dedicated to Lord Shiva. A suspension bridge here connects India and Nepal. The Kali River serves as a natural boundary between Nepal and Uttarakhand (India). The culture of Joljivi is a blend of Kumaoni and Nepali influences. The people share close relationships, celebrate fairs and festivals together, and have intermarriage ties. The road continues to Dharchula, located about 42 kilometers from Joljivi. Dharchula (970 m) also acts as a border town between Nepal and Uttarakhand, with the Kali River dividing the two regions. The riverbank is bustling with urban settlements on both sides, connected by three suspension bridges. The town appeared vibrant and unique.

Fig. 5.10 (**a**) Bangapani village. (**b**) Livestock and agricultural fields of Bangapani village. (**c**) Village ecosystem and the Gori River. (**d**) The Gori River joining with the Kali River

There are three famous valleys: Johar Valley in the upper catchment of the Gori River, Darma Valley in the upper catchment of the Dhauli Ganga (East), and Vyas Valley in the upper catchment of the Kali River (Singh and Rana 2002). From Dharchula, the steep climb began, and the road conditions were deplorable, with large stones scattered and parts of the road completely broken. Road construction work was ongoing throughout the area. Tawa Ghat, about 12 km from Dharchula, is located at the confluence of the two rivers—the Dhauli Ganga (East) and the Kali River. The route bifurcates at Tawa Ghat: one road leads to Gunji, Kuti, and Adi Kailash (Vyas Valley) along the Kali River, while another leads to Darma Valley and Panchachuli Mount along the Dhauli Ganga (East). Darma Valley is approximately 70 km from Tawa Ghat. The road condition along this entire 70 km stretch is the worst, with frequent bumps and the car jumping regularly. Occasionally, crossing a bridge provided some relief. A dam has been constructed on the Dhauli Ganga about 15 km from Tawa Ghat, creating a beautiful lake (Gautam 2019). After a certain point, the communication network becomes unavailable. The road follows steep slopes and is narrow, making it difficult for large vehicles to pass. Above 2000 m, the region remains snow-clad during the winter season. As we traveled toward Darma Valley, night fell, and the ascent became steeper, with patches of snow everywhere. The car skidded on icy patches, and mobile networks disappeared entirely. The setting sun cast a red glow on the snowy mountains, creating a scene as if the sun itself was spreading golden light. The author described the entire route during the journey in the following stanza.

> As darkness engulfed us, the Dhauli Ganga revealed its ferocious side, intimidating us. The road continued onward, passing through various villages such as Kanjyo (1540 m), Dar, Wongling, Seela, Chal, Nanking (trekking to Nag Parvat), Baling (trekking to Bayaksi Glacier), Bon, Filum, Gau (trekking to Sigla Pass and Adi Kailash), Saun, Dantu (trekking to Gabbey Bugyal), Dantu (trekking to Panchachuli), Danker (army check post), Tidang, Marchha, and Seepu (trekking to Mahadev Gufa, Asari Tal, and Moolti Bai). The entire trail is known as the Darma Valley of Dharchula Taluk. A road near Kanjyo crosses the river and leads to Narayani Ashram, about 40 km away. Near Kanjyo, a massive rock blocked the road, forcing us to take a detour. The road continued to climb, surpassing an altitude of 3000 m. Darkness surrounded us, with snow everywhere. Occasionally, a light piercing through the darkness startled us. The destination was still out of reach. Finally, we arrived at the place, where a few cottages and homestays were located. The cottages were quite expensive, costing ₹4000 per night, including breakfast and dinner. Surrounded by snow, the cold was intense.

The villages along the road or the course of the Dhauli Ganga practice agricultural and livestock farming for their livelihoods. Since the arable land is undulating and steep, soil fertility is limited, resulting in low crop production and productivity. However, the output from livestock farming is considerable. In the lower altitudes, people rear cattle—cows, bullocks, and buffaloes—while in the higher altitudes, they raise goats, sheep, lambs, and horses. Yaks are the main animals of the region, particularly at higher elevations. Milk production is high, so milk products—such as curd, butter, buttermilk, and ghee—are easily available in the villages and service centers. Apples, rajma, and potatoes are predominantly grown in the higher altitudes, which substantiate rural livelihoods. Settlements above 2500 m in the Darma Valley remain snow-capped due to heavy snowfall during the winter season. As a result, people migrate seasonally to the Dharchula and Joljivi areas and return to their summer dwellings during the warmer months. Panchachuli (Fig. 5.11), the highest peak in the Darma Valley, is a group of five mountain peaks, with its base camp located at Dartu village. The road is being widened up to Dartu village, and after Dartu, an Indian Tibet Border Police Camp (ITBP) is located, where several personnel reside. The area is barren during the winter, but in the summer, it is covered with flowers and medicinal plants, known as Bugyal. The tree line typically ends above 2800 m, and the snow line begins above 3400 m, although this can vary between winter and summer. Tourists visit the Darma Valley year-round, except during January, February, and March, when the region remains snow-capped and the climate is very harsh.

The same route returns to Tawa Ghat, taking almost 5 h to cover the distance. From Tawa Ghat, another route leads to Vyas Valley, famous for Adi Kailash, Om Parvat, and Parvati Kund—three highly revered religious sites for Hindus. Two villages in Vyas Valley, Kuti and Gunji, are particularly famous. While it used to be a challenging trek of about 50 km, these villages are now connected by an all-weather road. This route is also part of the famous highland pilgrimage to Kailash Mansarovar,

Fig. 5.11 (a) Panchachuli base camp Dartu. (b) Yak, an important highland animal is reared in Panchachuli area

sacred to both Hindus and Buddhists. Two strategic locations, Kalapani and Lipulekh, are situated in the extreme northwest part of Pithoragarh district (Kali Kumaon), on the borders of Nepal and Tibet. This region boasts rich cultural and natural diversity. The Shauka tribes inhabit the highlands of both Darma and Vyas Valleys. Tourists and pilgrims can visit Vyas Valley only with permission from the local government of Pithoragarh and the Indian Army, and are allowed access only during the summer and monsoon seasons. The area remains closed during winter.

Other notable places in the Kali Kumaon region include Loha Ghat, Champawat, and Tanakpur, some of which are located in plains, while others are situated in mountainous areas. The Kali River, the major river system in the region, holds immense natural, economic, environmental, and cultural significance. The journey from Kedarkhand to Manaskhand and Kali Kumaon spanned approximately 1400 km, passing through various districts and major river basins. The route traverses diverse landscapes—ranging from natural to cultural and forested areas—and crosses many river valleys, including the Ganga, Alaknanda, Pindar, Gomati, Saryu, Ramganga (E), Gori Ganga, Kali Nadi, and Dhauli Ganga (E). Along the entire route, the climate varies, transitioning from subtropical to temperate, cold, and frigid. At the final stop, temperatures dropped as low as −12 °C. Sub-cultural regions also differed significantly, including Devprayag, Rath, Chandpur, Badhan, Danpur, Thal, and Kali Kumaon. As a result, the food, beverages, attire, and languages varied accordingly. The journey through Manaskhand is a once-in-a-lifetime experience—a true blessing. Above 3400 m, the climate becomes harsh and uncomfortable, yet it also offers a sense of divine energy.

5.6 Conclusions

Like many cultural entities in the Garhwal Himalayas—such as the Panch Badris, Panch Kedars, Panch Prayags, and Char Dhams—the Kumaon Himalayas also boast three famous cultural valleys: Johar Valley, Darma Valley, and Vyas Valley. The Johar Valley, located at the source of the Gori Ganga (Milam Glacier), is home to the renowned town of Munsiyari in Uttarakhand. The Darma Valley lies at the source of the Dhauli Ganga, while the Vyas Valley is situated at the source of the Kali Ganga, and is famous for the revered sites of Adi Kailash and Om Parvat. All three valleys are located within the Pithoragarh district, a region known as Kali Kumaon. Furthermore, Jageshwar, Sun Temple in Katarmal, and Bageshwar are the most reverent cultural centers of Manaskhand.

References

Bhattacharyya HP (1922) The Skanda Purana: its text and its influence. Calcutta University Press, Kolkata

Bisht S (2016) Climate change and the Kumaon Himalayas: impacts and adaptation strategies. Natraj Publishers, Dehradun

Bisht BS, Dangwal LR (2022) Community participation in tourism development: a case study from Munsiyari, Kumaon Himalaya. J Ecotour 21(3):290–304

Doniger W (1991) Hindu myths: a sourcebook translated from the Sanskrit. Penguin Classics, London

Eliade M (1987) The encyclopaedia of religion, vol 9. Macmillan, New York

Gautam R (2019) Water resources and Management in Kumaon Himalayas. Indian National Science Academy, New Delhi

Gupta PC (2014) Economic diversification in the Kumaon Hills. Mittal Publications, New Delhi

Joshi KD, Bhatt A, Rana RS (2022) Changing cropping patterns and implications for food security in Kumaon Himalaya. J Mt Sci 19(5):1201–1214

Kashyap S (2018) Sustainable agriculture in the Kumaon Himalayas. Society for Himalayan Research, Dehradun

Keith AB (1917) The Skanda Purana, vol 3. Cambridge University Press, Cambridge

Mehta M, Tiwari PC (2021) Diversification of rural livelihoods and its impact on sustainability in Kumaon region, Uttarakhand. Int J Sust Dev World 28(6):565–577

Nair R (2005) The arts of Kumaon: traditional and contemporary practices. Natraj Publishers, Dehradun

Nautiyal S, Rao KS (2023) Eco-tourism and sustainable development in Kumaon Himalaya: a study of Nainital and Binsar. J Sustain Tour Stud 11(1):65–78

Negi GCS, Samal PK, Dhyani PP (2020) Sustainable livelihoods in the Indian Himalaya: integration of traditional and modern systems. Mt Res Dev 40(3):R12–R21

Pant GD (2010) Sustainable development in Kumaon: issues and challenges. National Institute of Rural Development, Dehradun

Pant MR, Pande A (2023) Climate-smart agriculture and policy responses in the central Himalayan region of Uttarakhand. Environ Dev 45:100765

Ramesh KV (1996) The Skanda Purana: a study in the development of Hinduism. Motilal Banarsidass Publishers, New Delhi

Rawal RS, Dhar U (2021) Traditional agricultural practices and agro-biodiversity conservation in the Kumaon Himalaya, India. Indian J Tradit Knowl 20(1):112–118. https://nopr.niscpr.res.in/handle/123456789/56587
Rawat SS (2007) The economy of Kumaon: agriculture, forests, and livelihoods. Indian Institute of Advanced Study, Shimla
Sati VP (2021) The Ganges: cultural, economic, and environmental significance. Springer, Cham, p 240
Sati VP (2022) Uttarakhand: society, culture, and pilgrimages. Today and Tomorrow, Printers and Publishers, New Delhi
Sati VP (2023) Sustainable forest management in the Himalaya. Springer, Cham
Sati VP (2024) Farming systems and sustainable agriculture in the Himalaya. Springer, Cham
Sati VP (2025) Exploring livelihood opportunities in the Kali Kumaon region of Uttarakhand Himalaya. Afr J Rural Dev 10(1):1–16
Sharma AK (2015) Environmental conservation and sustainable practices in Kumaon. Oxford University Press, New Delhi
Sharma R, Joshi V (2024) Migration and changing rural livelihoods in Kumaon Himalaya: a case of land abandonment and labour loss. Econ Polit Wkly 59(2):49–56
Singh RB, Rana PS (2002) Environmental management in the Kumaon Himalayas. Rawat Publications, Jaipur
Sood V (2003) Kumaoni culture: a study of social and religious practices. Kaveri Books, New Delhi
Tiwari S, Bhatt S (2021) Homestay tourism and socio-economic transformation in Uttarakhand Himalaya: a study of Kumaon region. Tour Recreat Res 46(4):512–523
Vogel J (2011) Religion and myth in the Kumaon Himalayas. Indological Book House, New Delhi

Chapter 6
Shiva and Vishnu: Unity and Continuity Through the Kali Gandaki River

Contents

Abstract Nepal, predominantly a Hindu country, is a mountainous nation with much of its territory covered by the Greater Himalaya. It is home to many Hindu temples, among which Pashupatinath—a Jyotirlinga—and Muktinath—a temple dedicated to Lord Vishnu—are especially prominent. A cultural route that begins in Kathmandu and ends at Muktinath passes through many rivers and ultimately follows the course of the Kali Gandaki River, a sacred river of Nepal. A large portion of arable land stretches along the terraces of these rivers, making Nepal an agriculturally dominated country. In addition, Nepal is renowned as a tourist destination, offering numerous cultural and natural attractions. Rich in abundant natural resources, Nepal's cultures, economies, and environments are explored in this chapter.

Keywords Pashupati Nath · Kali Gandaki · Kathmandu · Pokhara · Muktinath · Nepal

V. P. Sati, *Exploring the Mystical Himalaya*, Springer Geography, https://doi.org/10.1007/978-3-032-14275-7_6

6.1 Introduction

Nepal is a small, landlocked country nestled between two superpowers, India and China (Fig. 6.1). A significant portion of Nepal consists of the Himalayan Mountain range, known as the Nepal Himalaya, which is home to numerous snow-capped peaks. The highest peak in the Himalayas is located in the northern part of Nepal, along the border with China. Nepal's landscape also includes river valleys and fertile plains where traditional agricultural practices thrive. The country is interlaced with numerous rivers and their tributaries, creating rich alluvial plains. Some rivers hold cultural significance, such as the Kali Gandaki (Montmoli 2017). In its upper course, the river is revered as the Shaligram pilgrimage site, and along its banks lies the sacred Muktinath Temple (Zurick 2006). These rivers are vital to the livelihoods of local people, providing abundant water for both drinking and irrigation.

The people of Nepal predominantly follow Hinduism and Buddhism (Mitra 1882). These two religions share similar cultural practices, rituals, and customs, largely because Buddhism originated in India and spread to neighboring regions. In Indian belief, Lord Buddha, the founder of Buddhism, is regarded as the 12th incarnation of Lord Vishnu. Additionally, the Himalayas, which connect both India and Nepal, have a significant influence on the culture of both countries. Nepal is home to numerous highland and river valley pilgrimage sites that serve as cultural centers for both Hindus and Buddhists. For instance, Pashupati Nath in Kathmandu

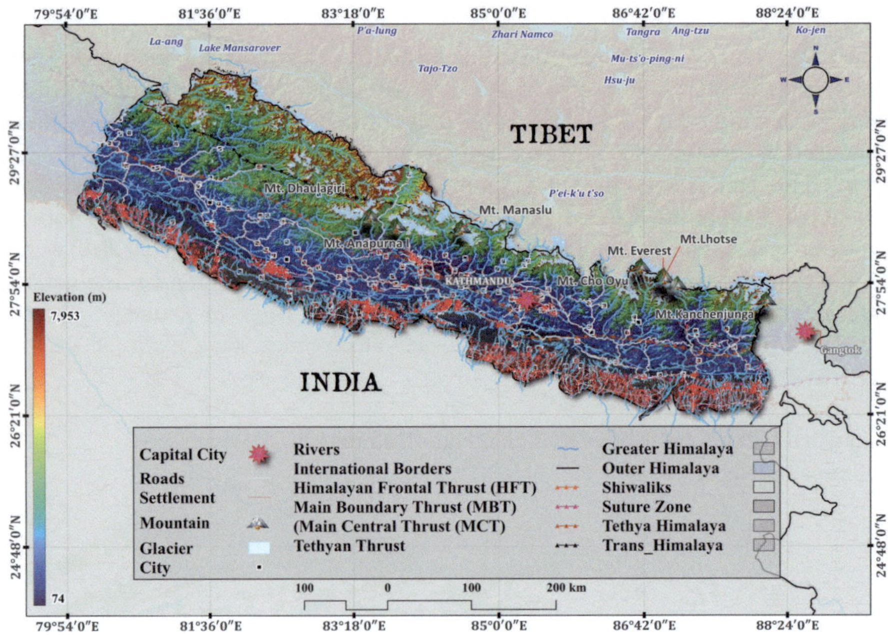

Fig. 6.1 Map of Nepal, also known as the Central Himalaya

Fig. 6.2 Lord Pashupati Nath Temple in Kathmandu. Photo By author

(Fig. 6.2) and Muktinath in Upper Mustang are revered as the abodes of Lord Shiva and Lord Vishnu, respectively. For Buddhists, Nepal also hosts many monasteries (Dubey 2023).

6.2 Agriculture, Tourism, and Economic Development

Tourism is one of the economic bases of Nepal, as it holds many destinations of natural and cultural interest. Nepal is the land of Lord Shiva and Vishnu. The temples of Pashupatinath in Kathmandu and Muktinath in Upper Mustang are world-famous. The country has many wildlife sanctuaries and national parks where eco-tourism is practiced (Sapkota 2023; Khanal and Gosai 2024). Adventure tourism has a wide scope, and Nepal is one of the key destinations for trekking and mountaineering. Recently, Nepal has been developing as a center for agri-tourism (Bhatta 2020; Khanal and Shrestha 2019). Sustainable tourism development may enhance Nepal's income and economy (Acharya 2023).

Nepal has an agrarian economy. It grows several crop races and cultivars along the altitudinal gradient. However, crop diversity varies according to altitude and climatic conditions. Most crops are grown during the monsoon season and are rain-fed. Only a small area is irrigated, mainly along river valleys and flood plains. Furthermore, modern innovations in agriculture are minimal, and crop production and productivity remain low—insufficient to meet the food needs of the population.

In contrast, food consumption is high due to the growing population in both mountain and plain regions (Karki et al. 2020).

Despite these challenges, Nepal has high potential for growing food grains, fruits, and vegetables due to its diverse and suitable climate. Sustainable agricultural intensification is essential to make Nepal self-sufficient in food production (Paudel et al. 2020). Nepal possesses rich agroecological practices, and farmers have vast traditional knowledge (Ojha et al. 2022). Recently, government policy initiatives have contributed to a certain extent of transformation in agriculture, with positive impacts on development (Upadhyay and Shivakoti 2021). Climate change has a substantial impact on agricultural practices across Nepal's altitudinal zones (Dahal et al. 2021). In response, the Government of Nepal has launched several adaptation practices to support smallholder farmers (Adhikari and Maharjan 2023).

Nepal's economic development has faced several challenges. Being a mountainous and landlocked country situated between two giant nations—India and China—poses significant geopolitical and logistical constraints. Its economy primarily relies on traditional crop cultivation and tourism (Bhattarai and Mishra 2023). Following the end of the monarchy and the adoption of democracy, the political landscape of Nepal has remained unstable, which has significantly affected the country's economy and overall development (Sharma and Upreti 2022). One of the major contributors to the Nepalese economy is remittances, as many educated citizens work abroad and send money back home (Pandey 2021). Foreign aid and tourism also serve as important sources of national income (Adhikari 2022).

Infrastructural facilities in Nepal are underdeveloped. The country has limited railway and air transport networks, and road conditions remain poor (Dhungel 2020). However, with financial assistance from India and China, infrastructure is gradually improving (Rai and Gautam 2023). China is supporting several infrastructure development projects, while India is primarily assisting in the energy sector (World Bank 2024). Other institutional sectors, such as banking and education, also remain in poor condition.

6.3 Kathmandu: A Land of Pashupati Nath

Kathmandu, the capital city of Nepal, is known for its beauty, culture, and educational significance. This scenic valley, surrounded by mountains and lush green grasslands, is home to Pashupati Nath, a revered jyotirlinga and manifestation of Lord Shiva, situated on the banks of the Bagmati River in Kathmandu (Fig. 6.3). The main temple houses a Shiva Lingam, displaying Shiva's head, including features such as the nose, mouth, eyes, and ear. Recognized as a UNESCO World Heritage Site, the Pashupati Nath Temple is a major cultural landmark. The Bagmati River, considered holy, runs beside the temple, where numerous smaller temples are also located within the premises. On special occasions, people take holy baths in the river, which serves as a primary cremation site for Hindus. They believe that cremating loved ones along the Bagmati's banks liberates the soul from the cycle of birth

Fig. 6.3 Front side of Lord Pashupati Nath Temple. Photo By author

and death, granting it Moksha (salvation). The main temple opens for Darshanam in the morning and evening, with Aarti rituals held during both times. Hundreds of pilgrims visit the temple every day.

Kathmandu is renowned as the "City of a Thousand Temples," with over 3000 temples adorning its landscape. The city showcases a harmonious blend of old and new migrant communities. Major festivals celebrated here include Ghode Jatra (a horse procession), Indra Jatra, Durga Puja, and Maha Shivaratri. Kathmandu's historical and religious sites are among its most captivating attractions. Although over 80% of the population practices Hinduism, the city's ethnic Buddhist communities play a vital role in preserving Buddhist culture. Numerous cultural centers, such as Changu Narayan Temple, Shivapuri Nagarjun National Park, and Patan Durbar Square, add to the city's cultural richness.

Most people in Nepal are non-vegetarian. Traditional dishes include Momos, Chatamari, Thukpa, Aloo Chop, Dal Bhat, Tingmo, Aloo Phingsha, and Juju Dhau. Nepali is the primary official language, spoken by nearly all Nepalese. Kathmandu serves as the center of Nepal's history, art, culture, and economy, boasting a multi-ethnic population. The region's spiritual landscape reflects influences from Kashmir Shaivism, the Nyingma School of Tibetan Buddhism, the works of Bhaktapur's Karmacharyas, and tantric traditions rooted in Hindu and Buddhist philosophies. Animal sacrifice is a common custom, with five male animals—buffalo, goats, sheep, chickens, and ducks—traditionally offered. Cows, however, are considered sacred and are worshiped. The Daura Suruwal is recognized as the national dress of Nepal.

Kathmandu was founded by Raja Gunakama Deva in 723 and was originally known as Manju-Patan (Shaha 1992). The current name, Kathmandu, is derived from a wooden temple, with "Kath" meaning wood and "Mandir" meaning temple. The Pashupati Nath Temple was constructed in 1596 by Raja Lachmina Singh. From 1768 to 2008, Kathmandu was ruled by the Shah family of the Gurkha people.

In 2024, the population of Kathmandu was 1,621,640, a significant increase from 104,479 in 1950 (UN World Urbanization Prospects). The Bagmati River flows through nearly the middle of Kathmandu, creating a broad valley, and meets the Vishnumati River near the city. Kathmandu's average altitude is 1400 m, and it is encircled by the scenic Chandragiri Hills, covered with temperate forests. Many temples and monasteries are situated atop these hills, enhancing their beauty.

In 2015, the Kathmandu Valley experienced a devastating earthquake that led to the loss of hundreds of thousands of lives. Part of the temple was damaged, and it took years for the city to restore the area, including the temple premises. Restoration efforts are still ongoing as of 2024. Nepal is an underdeveloped country with abundant natural resources that are not fully harnessed. The country has struggled to recover from the earthquake's impact for many years. Its economy remains largely agriculture-based, with low productivity, resulting in widespread food insecurity and malnutrition. Industrial development is hindered by challenging terrain, limited access to natural resources, and financial constraints.

6.4 Experiences From My Visit to Nepal

Nepal is renowned for its natural beauty, cultural richness, and educational opportunities. Every year, hundreds of seminars and conferences are organized across various cities, attracting academics from around the world. I have visited Nepal three times previously for academic purposes. This recent visit was for an international conference titled *Sustainable Mountain Development*, held in Pokhara, Nepal, from October 6–8, 2023. The organizers invited me to participate, and this was my first time visiting Pokhara. Known for its moderate climate, fertile agricultural land, temperate fruits, and large perennial lakes, Pokhara is a beautiful city situated in the Kali Gandaki River valley. It is also a key stop on the route to the renowned highland pilgrimage site Muktinath, also called the Shaligram pilgrimage, located in Upper Mustang. Considering Pokhara's natural and cultural significance, I gladly accepted the invitation. Although Nepal allows free entry for Indian nationals, a passport is still required. Immigration checks are mandatory at all international airports. At Kathmandu Airport, travelers must go through the passport control area. While Delhi Airport has undergone a tremendous transformation with world-class facilities, Kathmandu Airport remains small and congested.

6.5 Kathmandu to Pokhara

Traveling by air was the most convenient way to reach Pokhara from Aizawl, with flights to Kathmandu available from either Kolkata or Delhi, though flights from Kolkata are limited. I flew from Delhi to Kathmandu and stayed overnight in Kathmandu. The city offers a wide range of hotels, with Thamel being the best area to stay in. Thamel, a vibrant city center, is known for its diverse food and beverages, colorful clothing, and pub scene. The SAARC Secretariat and the former palace of the King are also located in the city center.

I stayed in Thamel, which made for a pleasant experience. Early the next morning, a bus to Pokhara was scheduled to depart from the Nepal Tourism Board (NTB) office, located about a 3 km walk from Thamel. I walked to the NTB office, where other conference participants were already waiting. After about half an hour, two comfortable Volvo buses arrived, and we began our journey to Pokhara. Although the distance between Kathmandu and Pokhara is only 250 km, the trip took around 12 h due to ongoing road widening projects, which left the road conditions in poor shape.

After passing through the Kathmandu Valley, we entered the Dhading District, which is part of the Trishuli River watershed. The road continued descending until we reached the Trishuli Valley, where the river flows through a deep gorge for a considerable distance. The altitude dropped to around 300 m. The entire Trishuli Valley is characterized by vast, irrigated arable lands, where crops like rice, maize, and pulses are cultivated. The landscape was truly spectacular, offering a scenic view throughout the journey. Our bus made a stop at a small Dhaba (restaurant) named Naubise, located along the route. A large crowd of fellow passengers was already gathered there for breakfast, which is locally referred to as "Khaja." We enjoyed a delicious breakfast, which included fried horse gram, Alu Chaap, Pakora, and Alu curry. After the break, we continued our journey, traveling along the winding road for nearly a hundred kilometers on the left side of the Trishuli River. The road was under construction in several sections, with some areas being particularly dangerous to navigate. Along the way, we passed by several service centers that cater to the surrounding rural areas. Notable service centers included Maleku, Damoli, Dulegaura, and Kotre, which provide essential food commodities, livelihood goods, and other necessities such as cosmetics to the local communities.

Many small perennial streams, fed by glaciers, flow into the Trishuli River from both banks. After traveling about 100 km, the Narayani River merges with the Trishuli River. We continued along the Trishuli River's course until it converged with the Gandaki River, also known as the Sapt Gandaki in Nepal (Fig. 6.4). The term "Sapt Gandaki" refers to the seven tributaries of the Gandaki River, which has a catchment area of 46,300 km^2, a significant portion of which lies within Nepal. The confluence of the Trishuli and Gandaki Rivers occurs at Muglin, a notable service center also referred to as Devasthanam, meaning “the place of God.” Here, we stopped for a lunch of delicious local dishes, with food and beverages available at very reasonable prices. From Muglin, the road splits—one branch heads toward Pokhara, while the other leads to Chitwan. Chitwan, a renowned national park and wildlife sanctuary in Nepal, is a popular destination for eco-tourists, both domestic and international. We took the road toward Pokhara. Along the way, we passed the major tributaries of the Gandaki River, namely the Madi and Saiti Rivers. The road runs in the opposite direction of the Gandaki River, continuing its journey further. Another prominent tributary of the Gandaki River is the Kali Gandaki (also known as Krishan Krishna Gandaki), which is considered a holy river in Nepal. The Kali

Fig. 6.4 The confluence of Trishuli with Sapt Gandaki River. Photo By author

Gandaki River also provides ideal locations for the construction of hydroelectric power projects. The Nepalese government has already developed a hydroelectric power project along this river, supplying electricity not only to the surrounding areas but also to the entire country.

The Kali Gandaki River is famous for its black Shaligrams, fossilized stones found mainly in the river's source area. Along the course of the river, hundreds of thousands of Shaligrams, both large and small, can be found. These Shaligrams are considered sacred, and the upper reaches of the Kali Gandaki River are regarded as a Shaligram pilgrimage site. The idols of Lord Rama and Goddess Sita, crafted from black Shaligrams, were brought from the Kali Gandaki River to the Ayodhya Ram Temple, where they are now enshrined in the Garbhagriha (sanctum). Pilgrims on their way to the highland pilgrimage site of Muktinath often collect black Shaligrams from the banks of the Kali Gandaki as part of their spiritual journey. The landscape throughout the route was nothing short of spectacular, with beautiful river valleys, expansive farmlands, lush forests, and the majestic snowy mountains adding to the region's allure. Despite the poor road conditions, the breathtaking views of the river valleys, agricultural landscapes, forests, and both rural and urban settlements were truly heartwarming, making the journey memorable.

From Muglin, the road gradually ascended as we made our way to Pokhara. Initially, the path was hilly, but as we progressed, the terrain became flatter, extending all the way to Pokhara. After a 12-h journey through the mountains, middle altitudes, and river valleys, we finally reached Pokhara. During this long journey, I observed that Nepal is agriculturally rich, with a remarkable agro-diversity. A wide variety of fruits, vegetables, and food grains are cultivated across the country. Paddy and maize are the primary crops, especially along the river valleys, while pulses and bananas are also commonly grown. The houses along the way are modern and beautiful, built with bricks and cement. Despite the scenic beauty, heavy rain accompanied us for most of the journey, creating a refreshing yet challenging atmosphere.

The journey between Kathmandu and Pokhara is often described as a divine path. While Kathmandu is regarded as the abode of Lord Shiva, Pokhara is known as the seat of Lord Vishnu. However, both Shiva and Vishnu are believed to reside together along the route. Upon reaching Pokhara, the rain continued to pour, setting a serene yet damp atmosphere. The city offered a range of high-quality hotels, and the food and beverages available were quite similar to Indian cuisine. We enjoyed a delightful gala dinner, which marked the beginning of the conference, which continued for 2 days.

Pokhara, also known as Lumhe, is situated at an altitude of about 700 m and is renowned for its heavy rainfall, particularly during the monsoon season, earning it the nickname "Cherrapunji of Nepal." The city boasts a rich blend of cultural and natural diversity (Holden and Sparrowhawk 2002). It serves as the gateway to the Annapurna Mountain Range, which is located close to the city, as well as the mythical Shangri-la. On clear days, the Annapurna Mountain Peak offers a breathtaking view from all corners of Pokhara, appearing so close it seems almost within reach, adding a panoramic charm that significantly enhances the city's beauty. Pokhara is a multifaceted city, known not only as the "City of Lakes" but also as the home of Goddess Maha Kali and Lord Shiva. It serves as the capital of Gandaki Province and is widely regarded as the tourism capital of Nepal.

The tranquil beauty of Pokhara has long been a source of inspiration for both tourists and writers. Its captivating environment, favorable climate, snowy peaks, serene lakes, and lush green surroundings have earned it the title of the "jewel in the Himalayas." The city is home to a cluster of nine lakes, with three—Phewa, Rupa, and Begnas—being especially popular (Nepal Tourism Board). This is why Pokhara is often referred to as the "City of Lakes" (Fig. 6.5). The climate in Pokhara is warmer compared to Kathmandu, with hot and humid summers and mild, pleasant winters. The city also boasts a beautiful airport, with flights operating to Kathmandu and other destinations, making it just a 30-min flight away from the capital. Roadways also provide an alternative means of travel. Over time, Pokhara has become a major tourism hub, attracting thousands of visitors each year. As of 2024, the population of Pokhara is estimated at 493,788, a sharp increase from just 4519 in 1950 (UN World Urbanization Prospects).

Pokhara was historically an important trade route connecting India and Tibet through the remote Himalayan region of Mustang. The area is home to several indigenous communities, including the Gurungs and Magars, who traditionally reside in the Mustang region. Another significant group, the Thakalis, inhabit the Thak Khola region of Mustang and are renowned for their entrepreneurial spirit. They are particularly known for running tea houses along the Annapurna trekking route. In recent years, Pokhara Valley has evolved into a hub for adventure tourism. The main adventure sports attracting tourists are paragliding and ultra-light aircraft flights, while other popular activities include boating, bird watching, trekking, and mountain biking.

Many academicians from around the world participated in the international conference, and the academic discussions were insightful and enriching. The food and beverages—served for breakfast, lunch, and dinner—were excellent. Nepali

Fig. 6.5 A panoramic view of Phewa Lake, located just in the middle of Pokhara city. Photo By author

cuisine, with its unique and flavorful dishes, stood out as a highlight of the event. The Mustang Valley is particularly renowned for its high-quality (organic) and abundant potato production. The variety of curry dishes made from Mustang Valley potatoes were diverse and exceptionally delicious, making them a must-try for any food enthusiast.

Pokhara is often referred to as the "City of Underground Water," and it is home to the largest perennial lake, Phewa Lake. This lake is not only beautiful but also surrounded by numerous temples dedicated to folk deities, attracting tourists who visit to worship these local gods. A long and scenic road runs along the bank of Phewa Lake, making it an ideal spot for morning and evening walks. Boating is a popular activity on the lake, providing a livelihood for many people involved in the boating industry. From almost every vantage point around the lake, the famous and sacred Annapurna Mountain Peak can be seen. In Hinduism, Annapurna is revered as Goddess Shakti, believed to provide nourishment and food to all people.

A perennial stream flows out of Phewa Lake, traversing the city and a nearby village, providing irrigation mainly for paddy fields. After flowing for over 3 km, the stream plunges into a deep ditch, creating a stunning waterfall. It then flows underground, disappearing completely and never resurfacing. This area is considered dangerous, as any mistake could result in a fatal accident. In fact, a tourist named David tragically slipped into the waterfall and lost his life, which is why it is now known as David Waterfall. On the opposite side of the ditch lies the Gupteshwar Mahadev Temple, dedicated to Lord Shiva. This temple is located approximately 200 m underground. A small trail leads into the cave, which is spacious enough to

Fig. 6.6 (a) Perennial stream inletting into below surface, making waterfall, opposite of the Gupteshwar Temple. It disappears from the surface and flows underground, (b) Gupteshwar Temple, 200 m below the surface in a wide cave where temples of Lord Shiva, Goddess Shakti, and Lord Vishnu lie. It is a holy place. Photo By author

house idols of various folk deities, including Lord Shiva, Goddess Shakti, and Lord Vishnu. The waterfall from the other side of the ditch and the cave converges, allowing visitors to experience the powerful mist from the waterfall (Fig. 6.6). At the entrance to the cave stands a statue of Lord Hanuman, where pilgrims first stop to worship before entering. Outside the cave, there is a large open area with a big statue of sleeping Vishnu, situated in a large water pond. On the other side of the road is a temple dedicated to Lord Shiva. Additionally, within Pokhara city, there is another glacier-fed perennial stream that flows underground, continuing as an underground stream.

The Kali Gandaki River flows through the Pokhara Valley, enriching the agricultural land and making it highly fertile. It irrigates vast expanses of arable land, supporting a diverse range of crops. The region is known for its rich agro-biodiversity, with numerous crop races and cultivars being grown. The Pokhara Valley is particularly renowned for its variety of fruits, including both sub-tropical and temperate species, contributing to the area's agricultural wealth and diversity.

6.6 Muktinath: A Divine Shrine

Muktinath Temple is situated at an altitude of 3710 m in the Mustang district of Nepal, along the famous Annapurna Circuit Trek (Fig. 6.7). To the north lies the Tibetan Plateau. Located at the heart of the Annapurna Conservation Area, Muktinath is considered a sacred place for pilgrims. The temple is dedicated to Lord Vishnu, a revered Hindu deity, but it is also sacred to Buddhists, as it is believed that Buddhist Guru Rinpoche meditated here. The temple is surrounded by towering snowy mountains and is built in the traditional pagoda style. In front of the Muktinath Temple are two holy water ponds—Laxmi and Saraswati Konda. It is believed that

Fig. 6.7 Muktinath Temple a highland pilgrimage located in the upper Mustang, Nepal. Photo By author

taking a dip in these ponds' cleanses one of sins and negative karma. Additionally, there are 108 cow-faced spouts, through which holy water flows into the ponds, considered sacred in Hinduism. Pilgrims worship these ponds before taking a holy dip.

Muktinath, also known as Muktikshetra, meaning the "Place of Moksha" in Hindu mythology, is a revered spiritual site. Buddhists call it Chumig Gyatsa, which translates to the "Place of 108 Waterspouts." Located near Muktinath Temple is the Goddess Jwala Mai Temple, which houses three eternal flames within the Gompa. It is believed that Lord Vishnu attained salvation from the curse of Brinda at Muktinath, which is why the temple is also known as the "Lord of Moksha." Devotees believe that this sacred shrine rose on its own. Numerous monasteries and Gompas are situated around Muktinath Temple, enhancing the area's spiritual significance. Several fairs and festivals, including Rishitarpani, Ram Navami, and Vijay Dashami, are celebrated with great enthusiasm, attracting thousands of pilgrims who visit for the Lord Vishnu Darshan. In Buddhist tradition, Lord Vishnu is worshiped as Avalokiteshvara.

6.7 Pokhara to Muktinath

A trek to Muktinath from Pokhara via Jomsom (2723 m) is part of the renowned Annapurna Circuit. This trek offers an incredible experience of the natural, cultural, and spiritual landscapes of the region. The Jomsom-Muktinath trek is an easy and low-altitude route, making it an excellent way to explore the Annapurna region. Along the way, trekkers are treated to spectacular views of Mount Dhaulagiri and other towering Himalayan peaks. Kagbeni, a sacred site near Muktinath, is a popular highland pilgrimage destination. The ancient villages of the Mustang region, rich in history and beauty, add to the allure of the trek. Other notable spots, such as Poonhill and Ghorepani, offer breathtaking panoramic views of the Himalayas. On the trek, the hot spring at Tatopani provides a refreshing break for both tourists and pilgrims. The cultural landscape is further enriched by the Gurung, Thakali, and Magar communities, each with their own distinct and rich cultural heritage.

From Pokhara to Jomsom, travelers have the option to journey by air, road, or trekking. The flight along this route is particularly scenic, offering breathtaking aerial views of the landscape. The trekking route is equally spectacular, passing through a diverse range of landscapes, from lush green paddy fields and sub-tropical rainforests to deep valleys, alpine slopes, and Tibetan plateaus. Trekkers also pass through the Kali Gandaki River, which is considered one of the deepest gorges in the world. The higher altitudes are part of a rain shadow region, characterized by dry, desert-like terrain similar to the Tibetan plateau. The areas closer to Pokhara and the hills are fertile and rich in vegetation, while the path from Jomsom to Kagbeni and along the Kali Gandaki River becomes rugged, eventually transitioning into a semi-barren landscape as trekkers move further into the region.

The Muktinath trek follows the course of the Kali Gandaki River, which originates from the Tibetan Plateau and flows through Mustang. Revered as a holy river, the Kali Gandaki is surrounded by cremation sites at its confluences, reflecting its spiritual significance. The river flows between the towering Dhaulagiri and Annapurna Mountain ranges, creating the world's deepest river canyons, with depths reaching approximately 3000 m. This region, once an ancient trading route to Tibet, has now become a popular trekking destination. The Kali Gandaki River valley offers stunning landscapes, with rapids at its source area creating dramatic effects with high water and deep holes. Besides its cultural and tourism importance, the Kali Gandaki River has significant potential for hydropower generation. The 144 MW Kali Gandaki run-of-river project, which began in 2002, includes a concrete gravity dam and is an important source of electricity for the region.

The cultural significance of the Kali Gandaki River is profound, connecting Nepal with Tibet and historically serving as an essential trade and pilgrimage route for Nepal, Tibet, and India. The river flows through numerous world-renowned pilgrimage sites, ancient villages, monasteries, and temples, rich in cultural heritage. These cultural centers attract pilgrims and tourists seeking spiritual solace and the opportunity to experience local traditions and customs. Both Hindus and Buddhists hold the Kali Gandaki River sacred, and many religious ceremonies and rituals are conducted along its banks. Pilgrims often take a holy dip in the river, believing it purifies the soul and helps in attaining Moksha (liberation from the cycle of birth and death). The Kali Gandaki is considered as holy as the Ganges River in India and is known as Krishna Gandaki in Nepal.

The river is especially revered for its Shaligrams—sacred black stones found along its banks, which are worshiped by Hindus. The idols of Lord Ram and Goddess Sita were carved from black Shaligram stones sourced from the Kali Gandaki River for the Ram Temple in Ayodhya. The upper reaches of the Kali

Fig. 6.8 (**a**) The Kali Gandaki River flowing below Nilgiri Mountain in the upper Mustang. (**b**) the Kali Gandaki River is making a spectacular landscape—settlements and agricultural fields, flowing in the middle Mustang. Photo By author

Gandaki are considered a pilgrimage site for these stones, with Damodar Kund, a glacier-fed lake in the high-altitude region, being a significant source of Shaligrams from fossil beds.

Additionally, the Kali Gandaki is part of the Sapta Gandaki system, which includes seven tributaries: the Kali Gandaki, Trishuli, Budhi Gandaki, Marsyangdi, Madi, Seti Gandaki, and Daraundi. These rivers together form the Narayani River, which ultimately joins the Ganges in Uttar Pradesh, India, further linking this sacred river to the greater spiritual and cultural landscape of the region.

The natural beauty of the Kali Gandaki River is truly spectacular, offering a diverse range of landscapes and unique geological features (Fig. 6.8). As a major tributary of the Sapta Gandaki (also known as the Narayani River in Nepal), it flows through a variety of natural formations, including rapids, rivulets, waterfalls, deep gorges, meanders, terraces, and floodplains. One of its most famous features is the Gandaki Gorge, one of the deepest river canyons in the world, which creates a breathtaking and dramatic landscape.

The Kali Gandaki is also renowned for its fossil deposits, particularly Shaligrams, which are sacred stones in Hinduism. The river's biodiversity is rich, with diverse flora and fauna thriving in its catchment area. This natural wealth supports the local economy by providing fertile soil for agriculture, with adequate irrigation from the river, contributing to sustainable livelihoods for the communities along its banks.

Trekking along the Kali Gandaki is a popular activity, drawing visitors for its white-water adventures, cultural richness, and stunning natural beauty. Major urban settlements along the river include Baglung, Ridi, Devghat, Narayanghat, Harinagar, and Valmikinagar, which are vital hubs for local trade and tourism. The area opened to tourism in 1991 and has since become a prime destination for trekking, offering a unique blend of culture, adventure, and scenic views.

The catchment area of the Kali Gandaki River is currently facing significant challenges, including deforestation, pollution, and human encroachment, all of which contribute to the loss of biodiversity and the degradation of the surrounding environment. To ensure the long-term sustainability of the river's ecosystem, it is crucial to implement effective management and conservation strategies. This includes promoting widespread afforestation campaigns to restore vegetation, establishing comprehensive waste management systems to mitigate pollution, and enforcing strict regulations on human activities around the river. These efforts are essential to restore the ecological balance and conserve the environment for future generations.

The Mustang region is divided into two distinct parts: the upper (northern) region and the lower (southern) region. Prior to 1950, this region was accessible via Tibet, but after China's invasion of Tibet, this route was closed. The lower Mustang was later reopened for trekking, while the upper Mustang remains restricted. Kagbeni, a village located on the banks of the Kali Gandaki River, serves as the dividing point between the upper and lower Mustang (Clara 1995). This village is also a key stop on the Shaligram pilgrimage route, with Muktinath being its final destination. Muktinath is not only a significant religious site but is also known as a Bon sanctuary, home to the Jwala Mai (Mother Flame), a natural gas vent that produces a continuous flame.

6.8 Conclusions

This chapter depicts the sacred journey from Kathmandu, the land of Lord Shiva, to Muktinath, the abode of Lord Vishnu, along a divine path. The route passes through various river basins, including the Trishuli, Gandaki, and Kali Gandaki, as well as the towns of Pokhara and Jomsom. The entire region is a stunning blend of natural beauty and cultural diversity. The Kali Gandaki River, regarded as the holiest, is particularly significant, with the upper reaches of the river known as the Shaligram pilgrimage site. Black Shaligrams, found in the river's upper course, hold immense religious value for Hindus. The journey along this divine path offers a glimpse into three distinct landscapes. The lower part is lush and green, with fertile, arable land and rich forests; the middle section is a cold, barren desert; and the upper part is dominated by snow-capped peaks. This region is home to several world-renowned mountain peaks, including Annapurna, which hold profound cultural significance. Muktinath, a prominent highland pilgrimage site and the seat of Lord Vishnu, is accessible only by trekking. While the trekking route is not excessively difficult, it offers a rewarding experience, making the entire journey truly worth visiting.

References

Acharya CN (2023) Sustainable tourism perspective of wildlife conservation areas: A case study in Nepal. J Econ Concerns 14(1-2):62–82. https://doi.org/10.3126/tjec.v14i1-2.62315

Adhikari DR (2022) Tourism, remittance, and sustainable development in Nepal: challenges and prospects. Sustainability 14(4):2283. https://doi.org/10.3390/su14042283

Adhikari R, Maharjan KL (2023) Adaptation practices to climate change in agriculture by smallholder farmers in Nepal. Sustainability 15(4):3221. https://doi.org/10.3390/su15043221
Bhatta K (2020) What is agritourism in Nepal? J Tour Hosp Educ 10:170–178. https://doi.org/10.3126/jthe.v10i0.28766
Bhattarai K, Mishra R (2023) Economic development in Nepal: trends, policies, and future challenges. J South Asian Dev 18(1):25–47
Clara M (1995) The Last Forbidden Kingdom. Land of Tibetan Buddhism, Mustang
Dahal KR, Aryal JP, Bhandari H, Rahut DB (2021) Climate change impact and adaptation practices in agriculture: Evidence from far-western Nepal. Climate Risk Manage 32:100306. https://doi.org/10.1016/j.crm.2021.100306
Dhungel KR (2020) Economic transformation in Nepal: an overview of structural change and policy reforms. Nepalese J Dev Rural Stud 17(1):1–14
Dubey Y (2023) Pashupatinath temple: The celestial abode of Lord Shiva in Nepal. Amar Granth. Retrieved 22 June 2024
Holden A, Sparrowhawk J (2002) Understanding the motivations of ecotourists: the case of trekkers in Annapurna, Nepal. Int J Tour Res 4(6):435–446. https://doi.org/10.1002/jtr.402
Karki S, Schleenbecker R, Hamm U (2020) You are what you grow and eat: farming styles and their influences on food consumption in rural Nepal. Appetite 149:104619. https://doi.org/10.1016/j.appet.2020.104619
Khanal B, Gosai KR (2024) Evaluation of ecotourism practices: a comparative study of lowlands and mountains of Nepal. J Tour Himalayan Advent 6(1):58–71. https://doi.org/10.3126/jtha.v6i1.67392
Khanal S, Shrestha M (2019) Agro-tourism: prospects, importance, destinations and challenges in Nepal. Arch Agric Environ Sci 8(1):464–471. https://doi.org/10.26832/24566632.2019.0404013
Mitra R (1882) The Sanskrit Buddhist literature of Nepal. Asiatic Scoeity of Bengal, Calcutta
Montmoli C (2017) A geological journey through the deepest gorge on Earth: the Kali Gandaki valley section, west-central Nepal. ResearchGate.net. Retrieved 27 July 2021
Ojha HR, Ghimire S, Pain A, Nightingale A, Khatri DB, Dhungana H (2022) Agroecological practices and knowledge in Nepal: navigating change and resilience. J Peasant Stud 49(3):453–476
Pandey S (2021) Foreign aid, remittances, and economic growth in Nepal: a time-series analysis. South Asia Econ J 22(1):56–75
Paudel B, Shrestha A, Rai RK (2020) Sustainable agricultural intensification in Nepal: patterns, drivers, and barriers. Agric Syst 184:102897. https://doi.org/10.1016/j.agsy.2020.102897
Rai DB, Gautam A (2023) Infrastructure and regional development in Nepal: examining spatial disparities. Asia-Pac J Reg Sci 7:21–42. https://doi.org/10.1007/s41685-023-00261-5
Sapkota N (2023) Socio-economic and environmental impact of ecotourism in Nepal. Nepalese J Hosp Tour Manage 4(1):57–71. https://doi.org/10.3126/njhtm.v4i1.53318
Shaha R (1992) Ancient and Medieval Nepal. Manohar Publications, New Delhi
Sharma S, Upreti BR (2022) Political economy of federalism and development in Nepal. Contemp South Asia 30(2):173–189
Upadhyay B, Shivakoti GP (2021) Agricultural transformation and food security in Nepal: a policy perspective. Food Secur 13:1139–1153. https://doi.org/10.1007/s12571-021-01161-z
World Bank (2024) Nepal development update: investing in human capital for a resilient recovery. The World Bank Group. https://documents.worldbank.org
Zurick D (2006) Illustrated atlas of the Himalayas. University Press of Kentucky, Lexington, p 153

Chapter 7
Harmony in the Himalaya: Buddhism and Bhutan

Contents

Abstract A large proportion of Bhutan's population practices Buddhism, followed by Hinduism—mainly among the Nepali Hindu community. Bhutan is a landlocked country, rich in natural resources such as land, forests, and water. Its economy relies heavily on the cultivation of traditional cereals, though both crop production and productivity remain modest. A significant portion of the population lives below the poverty line. Tourism is another key contributor to economic development. Remarkably, Bhutan has one of the highest happiness indices in the world. The country is home to world-famous monasteries that preserve its rich cultural heritage. Bhutan's physical landscape is broadly divided into three regions: the high Himalaya, the middle Himalaya, and the valley plains known as the Duars. This chapter explores all these aspects.

Keywords Buddhism · Monasteries · Happiness index · Duars · Thimphu · Bhutan

V. P. Sati, *Exploring the Mystical Himalaya*, Springer Geography,
https://doi.org/10.1007/978-3-032-14275-7_7

7.1 Introduction

Bhutan, a small yet awe-inspiring country, stands as a remarkable testament to its extraordinary landforms. Nestled in the embrace of the mighty Himalayas, Bhutan boasts a rich culture and heritage that is deeply rooted in its unique landscape (Fig. 7.1). The cultural fabric of Bhutan is vibrant, as it houses many temples and monasteries of significant spiritual value. Among these, ten major temples hold particular spiritual importance: Kyichu Lhakhang, Jokhang Temple, Jambay Lhakhang, Taktsang Temple, Tango Monastery, Kurjey Lhakhang Monastery, Tamshing Lhakhang Monastery, Punakha Dzong, Gangteng Monastery, and Tashichho Dzong. These revered sites serve as vital centers for the pursuit of spirituality and salvation, where Bodh monks engage in deep meditation and preaching. Devout pilgrims, following Buddhism, visit these temples to engage in worship and spiritual practices. In addition to Bodh monasteries, Bhutan also has a few Hindu temples, showcasing the country's inclusive approach to spirituality.

Bhutan is the only country in the world where Buddhism is the dominant religion, flourishing with the full support of the government, which has successfully preserved its Buddhist cultural heritage. Thimphu, the capital and largest city of Bhutan, is a scenic city located in the expansive Thimphu Chhu (River) valley, surrounded by temperate forests and the towering Greater Himalayan ranges. Due to the prominence of Buddhism, Thimphu is celebrated as a city of culture and beauty. The Bhutanese are known for their peaceful communities, and tourism plays a

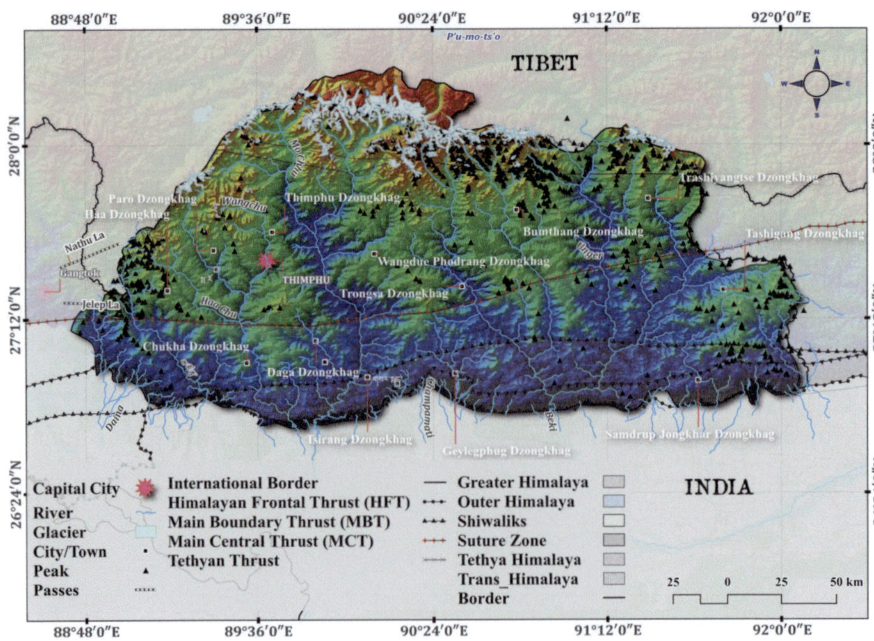

Fig. 7.1 The Kingdom of Bhutan, also known as the Bhutan Himalaya

crucial role in their economy, providing a primary source of income and livelihood. In the higher altitudes, they also rear animals, including yaks, goats, sheep, and lambs. Despite being a developing economy, Bhutan proudly boasts the highest happiness index in the world. India provides substantial financial assistance for Bhutan‘s economy and education, fostering a strong cultural relationship between the two countries.

Bhutan is a landlocked country, bordered by India on three sides—west, east, and south—with the Tibetan Plateau, the highest tableland in the world, defining its northern boundary. Bhutan is home to several impressive mountain peaks, with altitudes ranging from 5000 to 7000 m (Kaab 2005). Mount Everest, while not within Bhutan itself, lies close to its border. These majestic peaks remain snow-capped throughout the year, contributing to Bhutan‘s extensive glaciers. These glaciers serve as vital sources for the country's numerous perennial rivers, which flow throughout the year.

Bhutan was once a remote kingdom isolated from much of the world. However, in the second half of the twentieth century, it experienced transformative changes in transportation, primarily supported by India. Today, Bhutan is well-connected globally by air and has road and railway links to India. The country has a few fertile valleys in the Lesser Himalayas, where farmers actively practice agriculture. Thimphu, the capital, and Paro, home to Bhutan's only airport, serve as major centers for politics and commerce. Bhutan's terrain is defined by three main relief features: the Great Himalaya, the Lesser Himalaya, and the river valleys and Tarai region, known as the Duar plains.

7.2 Geography and Drainage

The Great Himalaya occupies the northern region of Bhutan, remaining snow-capped throughout the year. The highest altitudes exceed 7300 m, with high valleys situated between 3000 and 5500 m. Below these valleys are alpine pasturelands, which play an important role in grazing animals, including yaks, goats, sheep, and lambs. Pasturing is primarily practiced in the summer. The Tibetan Plateau in the north serves as a watershed, dividing the rivers that flow north and south. The climate in the north is dry and cold, often referred to as a cold desert. Before 1962, Bhutanese people living in the high Himalayan regions maintained trade relationships with Tibet, exchanging clothes, spices, and grains for salt, wool, and sometimes herds of yaks. The Himalayas hold strategic significance for Bhutan, providing a natural barrier that helps protect it from potential threats, including Chinese aggression.

The Lesser Himalaya, also known as the Inner Himalaya, forms the southern part of the Greater Himalaya in Bhutan. Areas on the windward side receive high moisture levels, attracting monsoon rains, which support temperate vegetation across much of this region. This natural vegetation enhances the panoramic beauty of the landscape and serves as a valuable source of income for the government. The valley region, encompassing areas like Paro, Punakha, Thimphu, and Ha, consists of flat land that receives moderate rainfall, making it ideal for sustainable agricultural practices. Consequently, a significant portion of Bhutan's population is concentrated in this zone, benefiting from the fertile land and favorable climate.

The Duars plain lies below the Lesser Himalaya and its foothills, stretching 12–16 km in width along Bhutan's southern border. This region consists of fertile valleys that receive abundant rainfall, ranging from 500 to 700 cm annually. As a result, the Duars are characterized by a hot and humid climate and are covered with dense, semitropical forests. The northern part of the Duars has a rough and rugged terrain. While small villages have been established in areas where forests have been cleared, most of the region remains densely forested, supporting a rich diversity of flora and fauna. Grasslands in the area are also cleared for rice cultivation, contributing to the agricultural productivity of the region.

Bhutan is home to several glacier-fed perennial rivers that flow southwards from the Great Himalaya, ultimately joining the Brahmaputra River system in India. These rivers, including the Torsa (also known as Amo), Wong (Raidak), Sankosh (Mo), and Manas, hold significant cultural and economic importance for the country. Bhutan's climate is diverse and influenced by various factors such as altitude, moisture, wind direction, and complex local variations. The country encompasses three major climatic zones: a hot, humid subtropical tract; a temperate zone; and an alpine tundra region. The climate in Thimphu tends to be generally warmer with high variability, while the Duars experience extreme heat, contrasting sharply with the extreme cold found in the northern regions of Bhutan.

7.3 Natural Vegetation

Bhutan boasts a rich diversity of flora that ranges from subtropical to temperate and alpine grasslands. This floral variety includes moist tropical deciduous trees, pine forests, coniferous species, and alpine shrubs. In terms of faunal species, some of the principal inhabitants include sambar deer, gaurs (wild ox), rhinoceroses, elephants, and tigers, including Bengal tigers. Dense forest areas, teeming with rich faunal species, are predominantly found along the Manas and Sankosh rivers in the central and eastern hill regions. Notably, the Royal Manas National Park and Jigme Dorji National Park serve as protected areas that are exceptionally rich in biodiversity, contributing to the conservation of Bhutan's unique wildlife and habitats.

7.4 Population and Ethnic Groups

The ethnic groups of Bhutan include the Bhutia (also known as Ngalop), the Nepalese, and the Sharchop. The Bhutia constitute the largest group, making up approximately half of the total population, and are predominantly found in the northern, central, and western regions. They are considered descendants of Tibetan immigrants who migrated from Tibet around the ninth century. The major languages spoken by the Bhutia are Tibeto-Burman and Dzongkha. The Gurung ethnic group is an integral part of the Nepalese community, which as a whole constitutes about one-third of Bhutan's population, primarily speaking the Nepali language. Both the Tibetan and Nepali groups face significant internal political challenges in Bhutan (van Driem 2001). The Sharchop, regarded as the hill tribes, are the earliest inhabitants of Bhutan and currently reside in eastern Bhutan, adjacent to the Indian states of Assam and Arunachal Pradesh.

A significant proportion of the population, approximately three-fourths, follow Buddhism, specifically the Tibetan variety. Buddhism was formerly the official state religion and was recently (in 2008) described as the spiritual heritage of the country. This tradition comprises three major branches: Tibetan Buddhism, Nyingma, and Kagyu. Nyingma, the oldest of these branches, has been present in both Bhutan and Tibet since the eighth century. Among these, the Drukpa Kagyu sect is the most prominent group in Bhutan, increasingly influencing the country's political and religious landscape (Tashi Kelzang 2023).

Bhutanese Buddhism possesses a unique character, with lamas (spiritual leaders) playing a vital role in both society and the preservation of Buddhist culture. These lamas engage in meditative disciplines and have attained the status of siddha, reflecting their spiritual accomplishments. Hinduism also holds significance, particularly within the Nepalese community, which constitutes approximately one-third of Bhutan‘s population. The Christian population is quite small.

The population in Bhutan is sparsely distributed due to the rugged terrain and fragile, precipitous landscape, resulting in a low number of inhabitants in many areas. The Greater Himalayan region is characterized by snow-capped peaks and frigid conditions, making it largely uninhabitable. In contrast, the Lesser Himalayan region, river valleys, and plains offer more favorable conditions for human settlement and agriculture, leading to a comparatively higher population concentration in these areas.

Historically, Bhutan had very few urban centers before the 1960s; however, significant transformations occurred thereafter, with rural areas evolving into urban centers due to economic development and road construction. As a result, more than 20% of the population now resides in urban areas. According to estimates for 2024, Bhutan‘s total population is approximately 793,200, representing only 0.01% of the world's population, and ranking 166th globally in terms of population. The population density is about 21 people per square kilometer across the country's area of 38,117 km^2. The national capital, Thimphu, has a population of 98,676 (Worldometer 2024).

7.5 Agriculture and Economic Development

The architecture of Bhutan exhibits significant variation from south to north. In southern Bhutan, the architectural style closely resembles that of neighboring India. In contrast, the high Himalayan region and the Lesser Himalayan valleys feature architecture that is typical of Tibetan influence. A prominent feature of Bhutanese settlements is the dzong, or fortress-monastery, which were originally constructed to protect communities from enemies. Today, dzongs serve as vital administrative centers as well as monasteries. Each densely populated area typically has one dzong, which functions as the main hub for political, economic, religious, and social life.

Phuntsholing and the Duars plains are particularly significant due to their large populations, infrastructural facilities, and agricultural production. Thimphu, located along the major highway, serves as the capital city. Additionally, Paro, situated approximately 50 km from Thimphu, is home to the country's only airport, which connects Bhutan to various South Asian countries, including Kolkata, Delhi, Dhaka, and Kathmandu, as well as to Bangkok.

Bhutan's economy is primarily based on traditional agriculture, livestock, tourism, and hydroelectricity. Agriculture is mostly rainfed, and farmers use organic manure in crop fields (Wangmo et al. 2023). The production and productivity of traditionally grown crops are low, and many farmers are marginal, resulting in food insecurity and malnutrition challenges. Recently, agriculture in Bhutan is undergoing a transition, with farmers shifting from subsistence agriculture to commercial farming (Tamang and Rai 2020), making it more resilient and productive (World Bank 2021). In contrast, the river valleys feature fertile plains with extensive arable land that benefits from irrigation (Statistical Yearbook of Bhutan 2019). However, due to climate change impacts in Bhutan's fragile Himalayan region, agriculture faces significant risks. Climate-smart agriculture is therefore the need of the hour, and the government is actively promoting it (Gyeltshen and Chhetri 2021). Furthermore, adopting sustainable agricultural practices—such as cultivating traditional crops and integrating modern innovations—may help achieve food security (Dorji et al. 2022).

Bhutan has an underdeveloped economy. Traditional agriculture, livestock farming, hydropower projects, and tourism are the major sources of income, economic activity, and livelihood (Asian Development Bank 2023). Economic diversification is considered a pathway to inclusive and sustainable growth (Dorji and Penjore 2021). Hydropower plays a significant role in Bhutan's economic development (Phuntsho and Wangdi 2020). The country has numerous glacier-fed perennial rivers with high potential for generating hydroelectricity. However, only a few projects are currently operational, and several are under construction.

At the same time, Bhutan's fragile landscape and undulating terrain necessitate careful environmental impact assessments of hydropower projects (Dorji and Fischer 2021). Bhutan is known for its harmonious society, and the Planning Commission of Bhutan (2021) is working to further enhance social harmony and sustainability. As a result, Bhutan ranks among the highest in the world in the Gross National Happiness Index.

Tourism development, on the other hand, holds tremendous potential, as Bhutan is endowed with snow-capped Himalayan mountain ranges, spectacular river valleys, panoramic forest landscapes, and beautiful agricultural fields. Promoting sustainable tourism can generate income and enhance employment opportunities (Wangchuk and Dorji 2022). Bhutan also has significant potential for eco-tourism development, given its numerous national parks and wildlife sanctuaries. The promotion of eco-tourism can serve as a pathway to sustainable development (Rinzin et al. 2007).

Although Bhutan possesses valuable mineral deposits, these resources are not optimally harvested. The main mineral resources include calcium carbide, limestone, dolomite, gypsum, coal, marble, quartzite, and talc, while lead, copper, tin, iron, and silver are considered minor resources. Additionally, Bhutan has significant potential for hydroelectricity generation, with the largest dam project being the Chhukha Dam, constructed on the Raidak River.

7.6 My Journey to Thimphu

On March 31, 2024, I, along with my wife, flew from Kolkata to Paro International Airport in Bhutan to participate in an international conference scheduled for April 1–2, 2024. We took a morning flight that lasted only 50 min, arriving in Paro around 10:00 AM. The journey was panoramic, offering stunning views of the Greater Himalayan peaks, with passengers on the left side of the plane treated to sights of Mount Everest and other majestic Himalayan summits. Paro Airport (Fig. 7.2), renowned as the world's smallest and toughest airport, is nestled in a deep valley between two giant mountain ranges. The flight makes several rounds before landing, but we touched down safely. Upon exiting the aircraft, we were greeted by pleasant weather. With two flights arriving within a 10-min interval, the airport was bustling with passengers. Given its small size, completing various formalities took us about an hour before we could exit.

At the airport, every passenger is required to pay INR 1200 per day as part of the Sustainable Development Fund (SDF). This fund is utilized by the Bhutanese government for tourism development and environmental conservation. After standing in a long queue to pay the SDF, we proceeded to immigration and finally made our way outside. The airport itself was beautiful, showcasing marvelous architecture that added to the charm of our arrival in Bhutan.

Paro (2200 m) is a picturesque town in western Bhutan, situated along the banks of the Paro Chhu (River). It is home to the country's only airport, which is located within Paro town. The Paro River creates a vast expanse of flat land interspersed with fertile agricultural patches, all surrounded by majestic mountains, resulting in a panoramic landscape. Paro serves as a business hub for Bhutan, connecting the nation to South Asia, and is recognized as one of the premier tourist destinations, featuring beautiful hotels and diverse cuisines. The town's buildings showcase exquisite architecture, contributing to its charm.

Fig. 7.2 Paro International Airport in Paro, Bhutan, is set against the stunning backdrop of the snow-capped Himalayan Mountain range. Photo By author

Before 1962, Paro was the main cultural, commercial, and political center of Bhutan until the national capital was relocated to Thimphu. Today, Paro functions as the summer capital of Bhutan. A notable landmark in the town is Fort Paro, a large rectangular seven-storied building that serves as the main cultural center. In addition to the airport, Paro is well-connected by the Indo-Bhutan National Highway. The average temperature in Paro ranges from 6 °C in January to 25 °C in July, with annual rainfall between 760 and 890 mm. Thanks to its fertile arable land, the Paro Valley is intensively cultivated, resulting in a densely populated river valley. Additionally, Paro is home to a beautiful monastery situated on the hillside to the left of the Paro River, further enhancing the town's scenic beauty.

The airport's unique location provides travelers with breathtaking views of the majestic peaks, creating a captivating first impression of Bhutan's natural beauty. The surrounding landscape is characterized by dramatic mountains, lush valleys, and the serene Paro River, making it a remarkable gateway to this picturesque country.

The Bhutan Tourism Department has implemented several rules for tourists arriving from outside the country. One essential requirement is that every tourist must have a tour guide and pay a fee of 1500 INR per day for their services. Additionally, hotel bookings and tour management must be arranged through authorized tour operators. Before our journey to Thimphu, we coordinated all the necessary arrangements with our tour guide, who was waiting for us outside the airport. Dressed in traditional attire and driving a Land Cruiser, he greeted us warmly in accordance with Bhutanese customs, offering both of us a white khata (a ceremonial scarf).

Once the formalities were complete, we set off on our journey to Thimphu, located 50 km from Paro. The road conditions were excellent, and we enjoyed the scenic drive. Notably, just a week prior, our Prime Minister, Shri Narendra Modi, had visited Thimphu, resulting in colorful banners and pamphlets adorning the roadside. We traveled along the right bank of the Paro River, and halfway to our destination, the Thimphu Chhu (where "Chhu" means river) joined the Paro River. This river continues its journey into Indian territory, ultimately merging with the mighty Brahmaputra River in Assam. Near the confluence of these two rivers, a small monastery is perched on a hillside, surrounded by terraced arable land (Fig. 7.3). A cement bridge allows us to cross the river where the Paro and Thimphu Chhu meet. After crossing, the road veered left, and we continued along the left bank of the Thimphu Chhu toward Thimphu City. Both slopes of the river were predominantly barren, but in areas where water was accessible, we spotted agricultural crops, lush grasslands, and temperate vegetation. Along the roadside, there were sparse settlements, a few restaurants, business avenues, and several schools, where we observed students along the way. The journey from Paro to Thimphu took us only about an hour.

This monastery, surrounded by tranquil landscapes, offers a peaceful retreat for visitors and serves as a focal point for spiritual practices in the region. The natural beauty of the area, combined with the spiritual significance of the monastery, creates a harmonious setting that reflects the rich cultural heritage of Bhutan. This was just the halfway point, and we continued on to Thimphu, covering the remaining distance. Thimphu, the capital city of Bhutan (Fig. 7.4), is situated in a long river valley along both banks of the Thimphu River. As we made our way to the core city area, we checked into the Shanti Deva Hotel, where we stayed for three nights. Conveniently located in the city center near the Clock Tower, the hotel was a great choice—neat, clean, and reasonably priced, with a nightly rate of 2000 ngultrum, including breakfast. After resting at the hotel, I ventured out for a walk along the city's main street. Most items available in the city, including food, clothing, and electronics, are imported from India, which makes their prices comparatively higher

Fig. 7.3 (**a**) The Paro River joins the Thimphu River at a picturesque confluence, where the serene environment is enhanced by the presence of (**b**) a monastery nestled on the hillside. Photo By author

Fig. 7.4 Panoramic view of Thimphu, the capital city of Bhutan, from Buddha statue. Photo By author

than those in India. Fortunately, most Bhutanese speak Hindi as a common language, and Indian cuisine is extremely popular. As a result, we encountered no issues with language or food during our stay.

The currency value of the Bhutanese ngultrum (nu) and the Indian Rupee (INR) was equal, and therefore, there was no problem with money exchange. In Bhutan, INR is widely accepted in all forms. During the entire journey, we used INR for all expenses. The streets were clean and wide, with well-maintained pedestrian ways. The cityscape was spectacular, with several beautiful squares and traditional architecture buildings, where tourists gather and enjoy themselves every day. The currency value of the Bhutanese ngultrum (nu) and the Indian Rupee (INR) is equal, making money exchange seamless for travelers. In Bhutan, the Indian Rupee is widely accepted in all forms, which allowed us to use INR for all our expenses throughout the journey. The streets of Thimphu were clean and spacious, featuring well-maintained pedestrian pathways that made walking enjoyable. The cityscape was spectacular, showcasing several beautiful squares and traditional architectural buildings. These vibrant spaces attract tourists daily, providing opportunities for relaxation and enjoyment as visitors immerse themselves in the local culture and scenery.

Thimphu, the capital city of Bhutan, offers a stunning panoramic view characterized by its unique blend of traditional architecture and natural beauty. Nestled in a lush river valley, the city is surrounded by majestic mountains that create a breathtaking backdrop. The Thimphu River winds through the valley, enhancing the landscape with its serene waters. The skyline is adorned with colorful buildings, dzongs (fortresses), and monasteries that reflect Bhutan‘s rich cultural heritage. Key

(a) (b)

Fig. 7.5 (**a**) The statue of Lord Buddha majestically overlooks the city of Thimphu from its hilltop perch, embodying peace and tranquility. This iconic landmark serves as a symbol of Bhutan's rich spiritual heritage. (**b**) The Goddess Panchayani Temple, a renowned Hindu temple, houses the idols of five female deities, including Maha Kali, Durga, Lakshmi, Saraswati, and Gayatri. This serene temple attracts numerous devotees and tourists, providing a space for prayer and reflection amid its tranquil surroundings. Photo By author

landmarks, such as the Tashichho Dzong and the Buddha Dordenma statue (Fig. 7.5), stand out prominently against the picturesque scenery. The vibrant streets bustle with activity, showcasing a harmonious mix of local life, shops, and cafes, while the clean, well-maintained roads contribute to the overall charm of the city. This panoramic view encapsulates the essence of Thimphu, where tradition and modernity coexist beautifully in the heart of the Himalaya.

In the evening, we took a leisurely stroll through the city center of Thimphu and along the banks of the Thimphu River. It was April, and the climate was pleasantly mild. Thimphu experiences extremely cold winters, often accompanied by snowfall, while the summers are quite enjoyable. The Thimphu River, a clean, glacier-fed perennial river, flows gracefully through the heart of the city, adding to its charm. The residents belong to diverse ethnic groups, yet they coexist peacefully. Their unique languages, food habits, and traditional clothing reflect their rich cultural heritage. Although many people face economic challenges, Bhutanese culture boasts the highest happiness index in the world. The community thrives in harmony, both among its members and with the surrounding nature. We enjoyed a nutritious and delicious dinner, bringing a satisfying end to a wonderful day.

Thimphu and its surrounding areas boast numerous cultural and natural destinations worth exploring. One of the most remarkable sites is the gigantic Buddha statue, located on the hills overlooking Thimphu City at an altitude of 2700 m, about 20 km from the city center. Standing at an impressive height of 52 m, this statue was constructed between 2006 and 2012 by the King of Bhutan to commemorate his 60th birthday. The expansive premises are enveloped by beautiful evergreen temperate forests, creating a tranquil atmosphere. A few shops at the site cater to tourists, offering snacks and drinks. We started our day early, eager to visit several attractions. Our first stop was the Buddha statue, which is situated in a serene environment that allows for stunning views of Thimphu City below. The entire area was breathtaking, and despite the chilly morning air, we were captivated by the beauty surrounding us. After spending about half an hour soaking in the views and the peaceful atmosphere, we began our descent back to the city.

Below the Buddha statue lies the Hindu Temple of Goddess Panchayani, dedicated to five female folk deities: Maha Kali, Durga, Lakshmi, Saraswati, and Gayatri. The temple is situated in a serene environment, set in a vast area that invites reflection and tranquility. Every day, tourists and pilgrims from Bhutan, Nepal, and India visit the temple to offer their prayers to these revered deities. The sound of chanting within the temple and its surroundings fosters a sense of peace and inner strength, enriching our experience. While Bhutan is predominantly home to monasteries, Hindu temples are few and typically managed by Nepali priests. The entire route leading to the temple is breathtaking, showcasing rich biodiversity, including vibrant floral and faunal species. Dense forests of pine thrive in the lower altitudes, while coniferous deodar trees flourish at higher elevations, creating a panoramic forest landscape. From the hilltops, visitors are treated to sweeping views of a long chain of snow-capped Himalayan peaks, further enhancing the area's natural beauty.

Thimphu, Bhutan, is home to two renowned higher education institutions: Royal Bhutan University, situated in the city center, and Royal Thimphu College, nestled within a dense forest area near the city. I had the privilege of delivering a lecture on tourism development in the Himalaya to the graduates at Royal Thimphu College, which is why we hurried to the campus immediately after our visit to the Hindu temple. The college is located approximately 30 km from the temple, surrounded by lush temperate forests and traversed by a glacial-fed perennial stream, creating a cool and refreshing climate on the premises. The college buildings, including institutional and residential facilities along with hostels, are beautifully designed and well-maintained, showcasing traditional Buddhist/Tibetan architecture.

I was invited to give the lecture by Professor Sanjeev Mehta from the Department of Management at Royal Thimphu College. My presentation was scheduled for 10:00 am in the conference hall. Upon arrival, I found the hall filled with eager management students. After a formal introduction and inaugural function, I delivered an engaging hour-long lecture. The students were welcoming, knowledgeable, and polite, making for a highly interactive session. Following a brief question-and-answer segment, the program concluded on a positive note. We then enjoyed a delicious lunch at the student canteen, featuring memos and noodles, before returning to our hotel in Thimphu city.

After enjoying our lunch, we had some time to explore more of Thimphu City. One of the highlights of our journey was visiting Bhutan‘s Parliament building (Fig. 7.6), which is beautifully situated on the banks of the Thimphu River in a wide river valley. The building's architecture is striking, characterized by red roofs of equal height, creating a harmonious and picturesque view. Bhutan operates under a democratic system, yet major decisions are primarily made by the King, reflecting a unique blend of tradition and modern governance. The core city area is vibrant, filled with numerous hotels and restaurants offering both continental and local cuisine. We also discovered a variety of handloom shops tucked away in the corners of

Fig. 7.6 The Bhutanese Parliament is situated on both banks of the Thimphu River, creating a scenic backdrop in the heart of the city. Photo By author

the city, featuring a range of clothes and gift items that showcase Bhutan‘s rich textile heritage.

As we continued our exploration, we noticed rural settlements perched atop the hills, offering breathtaking views of the mighty Himalayas. Surrounding Thimphu City, we observed patches of dense forests, areas of degraded land, terraced agricultural fields, and in the higher altitudes, the stunning sight of snow-capped mountain peaks. This picturesque landscape beautifully encapsulates Bhutan‘s natural beauty and cultural richness, making our visit even more memorable.

The building’s unique architecture reflects Bhutan‘s cultural heritage while serving as a vital center for the nation’s democratic processes. With the flowing river nearby, the Parliament area is surrounded by natural beauty, enhancing its significance as a governmental landmark. The design incorporates traditional Bhutanese elements, making it not only a functional space for governance but also an attraction for visitors interested in Bhutan‘s political landscape and cultural identity.

Everything was going smoothly. We enjoyed fast food and Chaat dishes paired with refreshing fruit juices—apple and orange. Several important tourist destinations remained on our itinerary, and one of them was Dorchulla Pass. Therefore, the next morning, we set out for Dorchulla Pass, located approximately 40 km from Thimphu City, at an altitude of 3130 m and surrounded by dense temperate forests featuring various species of rhododendron. It was the season for blooming rhododendrons, and the entire landscape was stunning, adorned with vibrant red flowers. After traversing picturesque river valleys and lush forests, we finally arrived at Dorchulla Pass. This location is both historically and naturally significant. Encircled by two mountain ridges, Dorchulla Pass creates a panoramic open space. In 2003,

Fig. 7.7 Dorchulla pass, a beautiful square and stupas. Photo By author

the King of Bhutan constructed 108 stupas in memory of the Bhutanese soldiers who lost their lives in a conflict with militants from the south. These white stupas were built on the slopes of the hilltops, encircled by a winding road. On the opposite side, atop Dorchulla Pass (Fig. 7.7), stands a Buddha Temple (monastery), housing idols of various gods and goddesses, including Lord Shiva. Bhutanese people believe that Lord Shiva is the younger brother of Buddha. We took a moment to worship Lord Shiva. Although the Bhutanese Government had established a fee to visit the Buddha Temple, the official kindly exempted us from the charge due to the brotherly relations between Bhutan and India. We spent about half an hour in the monastery before beginning our descent.

There were several tea stalls and restaurants serving food and beverages to tourists, providing a vibrant atmosphere where one could enjoy the breathtaking views of the long Himalayan ranges. With hearts full of joy, we returned to Thimphu City. Before heading to our hotel, we stopped at a viewpoint that offered stunning sights of various buildings and monuments showcasing marvelous architecture. In the city center, there is a beautiful stupa that is open to visitors from 9:00 AM to 5:00 PM, with an entry fee of 500 INR. However, access is free before 9:00 AM and after 5:00 PM. The dinner we had that evening was delicious, and we settled in for a restful night's sleep. The next day, we strolled through the city, shopping for a few souvenirs. We relished the local cuisine and the pleasant climate, which invigorated us and enriched our experience. After thoroughly exploring Bhutan City, we returned to Paro for an overnight stay. We enjoyed a leisurely walk through the Paro market and finally, on April fourth, we boarded our flight back to Kolkata, cherishing the memories of our wonderful trip.

7.7 Conclusions

This chapter describes the cultural, social, economic, and natural landscapes of the Bhutan Himalaya. The study further reveals that, although Bhutan is a relatively poor country, it boasts a high happiness index. The Himalayas provide significant resources to Bhutan, including water, forests, and fresh air, fostering harmony between nature and culture. This majestic range draws a line connecting the Himalayan nations into a cohesive entity. Bhutan is also renowned for the flourishing of Buddhism; indeed, if Buddhism exists in its entirety anywhere, it is in Bhutan. The Himalaya instil a sense of harmony in the Bhutanese people, allowing them to thrive in a peaceful atmosphere (Sati 2024). Bhutan's physical landscape is profoundly shaped by the Himalayas, featuring three primary topographies: the snow-capped Greater Himalaya, the Lesser Himalaya, and the river valleys and Tarai regions (the Duars). These distinct topographies significantly influence population distribution, with no population residing in the Greater Himalaya, a moderate population in the Lesser Himalaya, and a high population in the river valleys and Tarai regions. This variation in population density is largely attributed to climatic conditions and the nature of the terrain. Consequently, cultural practices also differ from one location to another, reflecting the changes in climate and topography. The economic development of Bhutan remains limited, with most people living on moderate to low incomes and facing restricted livelihood options. Agriculture serves as the primary source of income for many communities, with livestock rearing—particularly goats, sheep, lambs, and cattle—also playing a significant role in the economy. Additionally, tourism has emerged as a vital source of state revenue, with considerable future potential due to Bhutan's rich array of cultural and natural attractions.

References

Asian Development Bank (2023) Bhutan: economy profile and outlook 2023. Asian Development Bank, Manila

Dorji L, Fischer TB (2021) Strategic environmental assessment of hydropower development in Bhutan: prospects and challenges. Energy Sustain Soc 11(1):1–12. https://doi.org/10.1186/s13705-021-00297-9

Dorji T, Penjore D (2021) Economic diversification in Bhutan: a pathway to inclusive and sustainable growth. J Bhutan Stud 45(2):55–74

Dorji T, Wangdi J, Tamang B (2022) Sustainable agricultural practices in Bhutan: challenges and opportunities. J Bhutan Stud 46(1):23–41

Gyeltshen T, Chhetri N (2021) Climate-smart agriculture in Bhutan: a policy perspective. Bhutan J Agric 4(1):15–30

Kaab A (2005) Combination of SRTM3 and repeat ASTER data for deriving alpine glacier flow velocities in the Bhutan Himalaya. Remote Sens Environ 94(4):463–474

Phuntsho T, Wangdi J (2020) Hydropower and economic development in Bhutan: an assessment of sustainability and trade-offs. South Asian Econ J 21(1):76–94

Planning Commission of Bhutan (2021) Twelfth five year plan (2018–2023): a path towards just, harmonious and sustainable society. GNHC Secretariat, Thimphu

Rinzin C, Vermeulen WJV, Glasbergen P (2007) Ecotourism as a mechanism for sustainable development: the case of Bhutan. Environ Sci 4(2):109–125. https://doi.org/10.1080/15693430701365420

Sati VP (2024) Thimphu: a city of rich culture and beauty. https://medium.com/@sati.vp/thimphu-a-city-of-rich-culture-and-beauty-661cf90dab1f

Statistical Yearbook of Bhutan (2019) Thimphu: National Statistical Bureau

Tamang B, Rai N (2020) Agricultural transition in Bhutan: moving from subsistence to commercial farming. Bhutan Dev Rep 5(2):89–103

Tashi Kelzang T (2023) World of worldly gods: the persistence and transformation of shamanic bon in Buddhist Bhutan. Oxford University Press, Oxford, p 296

van Driem G (2001) Languages of the Himalayas: an ethnolinguistic handbook of the greater Himalayan region. Leiden, Brill, p 915

Wangchuk S, Dorji Y (2022) Tourism, employment, and economic resilience in Bhutan post-COVID-19. Bhutan J Dev Policy 6(1):22–39

Wangmo T, Dorji T, Penjor D (2023) Organic farming and food security in Bhutan: assessing outcomes and limitations. Asian J Agric Dev 20(1):67–84

World Bank (2021) Bhutan agriculture sector review: towards a more resilient and productive agriculture. The World Bank, Washington. https://documents.worldbank.org/en/publication/documents-reports/documentdetail /483861621276343196/

Worldometer (2024) Bhutan population. https://www.worldometers.info/world-population/bhutan-population. Assessed 28 October 2024

Chapter 8
Beyond Borders and Culture

Contents

Abstract Tibet was an independent nation before 1956, after which it was annexed by China and is now a part of the country. Known as the highest plateau in the world (above 4000 m in altitude), Tibet is a cold desert. Winters remain frigid, and summers have a harsh climate. The region is characterized by barren, stony land with snow-capped peaks. Tibet's economy is primarily based on livestock rearing, with yak being the main animal. Goats, sheep, and lambs are also important, providing milk, meat, and wool. Tibetans follow Buddhism and possess a rich cultural heritage. The region is home to many world-famous monasteries of great religious and historical significance. In this chapter, the history, culture, economy, and climate of Tibet are elaborated upon.

Keywords Cold desert · Lhasa · Yak · Buddhism · Mt. Gonga · Tibet plateau

8.1 Introduction

Tibet, located beyond the main range of the Himalayas, is often referred to as the "Trans-Himalaya" (Fig. 8.1). This elevated plateau, situated at an altitude exceeding 4000 m, is famously known as the "Roof of the World" and the "Third Pole" (Yao et al. 2012). Characterized as a cold desert, Tibet experiences extremely harsh

V. P. Sati, *Exploring the Mystical Himalaya*, Springer Geography,
https://doi.org/10.1007/978-3-032-14275-7_8

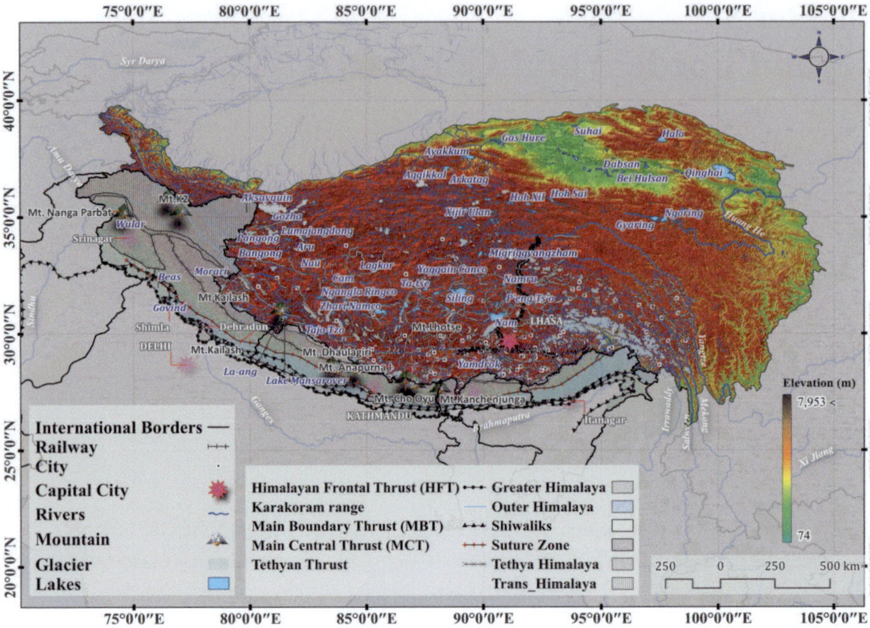

Fig. 8.1 The Tibet highland, a Trans-Himalaya

winters lasting about 6 months during which the land remains frozen and life is particularly challenging. In contrast, summers are mild, making it an ideal time for tourism. Known as the land of a thousand snow-fed streams and rivers, Tibet‘s terrain is largely barren and rocky. However, patches of greenery, including sparse trees and grasslands, are visible along riverbanks and stream corridors. While the winters are severe, the summer season brings a pleasant change, enhancing the region’s natural appeal for visitors (Yang and Zheng 2004).

The Tibetan people belong to the Mongoloid race. Their primary occupation is animal husbandry, and they also engage in the trade of gems and jewelry. Yak, goats, sheep, lambs, horses, and dogs are among the main animals they rear. Besides these livelihoods, Tibetans also practice limited farming, cultivating crops like potatoes and maize, along with some medicinal plants. However, life in Tibet remains challenging, as many struggle daily to secure adequate meals (Yeh and Henderson, 2008). Historically, Tibet was an independent country until 1956 (Goldstein 1991), with prosperous trade relations with India, Nepal, and Bhutan. Today, it is an autonomous region within China.

Climatic conditions in Tibet are very harsh. Over time, the Tibetan Plateau has been warming and influencing the precipitation patterns of South and East Asia (Wang et al. 2008). Over the last millennium, moisture changes in the northeastern Tibetan Plateau have led to a drier climate (Chen et al. 2010). Furthermore, snow cover has decreased and glaciers are melting, altering the cryosphere of the Tibetan Plateau (Kang et al. 2010). Small-scale agriculture and animal husbandry are the main occupations of Tibetans. Recently, farmers and herders have been struggling to cope with environmental and social changes (Qiu 2016). Tibetan herders' perceptions of grassland degradation and climate change in the eastern Tibetan Plateau support the evidence of climate change (Cao et al. 2013). This region has experienced tremendous rangeland degradation in recent decades (Harris 2010).

The agropastoral tradition in Tibet is unique and centuries old. It is practiced in the highlands during the summer. In winter, the pasturelands remain closed due to snow (Wu et al. 2015). In recent years, Tibetan grasslands have been depopulating due to climate change and a decrease in grazing animals (Foggin 2008). Factors such as urbanization and modernization are contributing to the decline of pastoralism in Tibet (Fischer 2014). This trend is continuing and is leading to the end of nomadic pastoralism (Næss 2013). Infrastructure development is also progressing, contributing to economic transformation (Fischer 2014; Wang and Zhou 2020).

Tibet has a rich cultural heritage. The monasteries in Lhasa and other parts of Tibet are world-famous, attracting hundreds of thousands of pilgrims every year (Nyima and Williams 2022). It is a paradise on Earth, with numerous natural locales—truly a playground for tourists (Hillman 2003). Due to its remoteness, a large portion of the population remains poor. Tourism development can help boost the local economy (Zhao and Ritchie 2007).

I have completed over ten international academic visits to various regions of China, with a primary focus on Chengdu, the capital of Sichuan Province in southwestern China, which borders the Tibetan Autonomous Region. The Tibetans residing in the southern part of Sichuan Province, mainly around the Maoshan Mountain area, form a minority community. The Chinese government has provided them with special privileges to support their well-being. During two separate visits to different parts of Tibet in 2016, I had the opportunity to visit Tibetan homes and engage with the local people. These experiences allowed me to study Tibetan land, culture, and communities in greater depth.

8.2 A Journey to Chengdu via Lhasa

It was August 31, 2016, and the rain poured steadily throughout the day. Kathmandu airport was bustling and overcrowded, with its limited space struggling to accommodate the passengers. My Eastern Chinese Airlines flight faced a 2-h delay, making me anxious about arriving on time in Chengdu, where I had informed Prof. Deng Wei of my arrival schedule. He mentioned that one of his Ph.D. scholars would meet me at the airport. My journey began with a flight from Kathmandu to Lhasa, a city often referred to as the "forbidden city" and situated above Mount Everest amidst a landscape of snow-capped peaks. As we flew over the Himalayas, the pilot announced that we were passing above Mount Everest (Fig. 8.2). This sparked excitement among the passengers, all eager for a glimpse of the world's highest mountain. Only those seated on the side facing Everest had the chance to witness its grandeur. Fortunately, I was among them, able to capture this

Fig. 8.2 Flying over the Himalaya between Kathmandu and Lhasa in the middle of agglomeration of mountain peaks is the Mount Everest. Photo By author

breathtaking sight in photographs and videos, preserving the memory of viewing the world's tallest peak.

After flying over the vast expanse of the Himalayan range, our flight finally touched down at Lhasa International Airport. Nestled in a valley, the airport lay beside the snow-fed Lhasa River, a tributary of the Tsangpo (Brahmaputra) River. Security at the airport was notably strict, with rows of fighter planes lined up on a large open ground, seemingly ready for combat. Although my flight was continuing to Chengdu, the airport officials required us to de-board with all our luggage for additional formalities. Once the formalities were completed, we re-boarded the flight to Chengdu. I found myself seated next to a Buddhist woman who appeared eager to communicate, though a language barrier stood between us. She made an attempt by saying "hun…," and from her expression, I guessed she was asking my name. I replied, "Indu," to which she responded with a smile, saying "Buddha, me Buddha." I smiled back and said, "Happy to meet you." She laughed heartily, sharing a warm moment despite the language gap.

Tibetans share a deep affinity with Indians, largely due to their shared connection with Lord Buddha. Buddhism, which originated and flourished in India, was introduced to Tibet, establishing a strong cultural bond between the two regions. Another link is the Himalayas, a majestic range that spans both countries. For Tibetans, the Himalayas are referred to as "Shimalaya," symbolizing Lord Shiva, and this spiritual association fosters a profound respect for their Indian counterparts. Buddhism has had a lasting impact on Tibet, making it home to numerous renowned monasteries. Alongside their spiritual traditions, Tibetans uphold unique cultures and customs that distinguish their heritage and way of life.

Lhasa, often called the "forbidden city," is situated at an altitude of 4000 m and is renowned for its many monasteries. As the capital of the Tibet Autonomous Region in southwestern China, Lhasa is revered as a holy land by Tibetan Buddhists for its deep historical and cultural significance. The city historically served as the religious center of Tibet, with much of its population comprising Buddhist monks and lay followers. Its sacred status and cultural heritage continue to draw pilgrims and visitors from around the world. The landscape of Lhasa is distinctive, featuring a tableland interspersed with small hills. The surrounding terrain is mostly barren, with occasional green patches where water sources are present. The climate is cold and dry, with harsh winters that blanket the entire region in snow and make it exceptionally frigid. Summers, however, bring milder, pleasant weather, making this season ideal for visits. As a result, pilgrims and tourists commonly choose to visit Lhasa during the summer months.

The four-story Jokhang Temple, located in the heart of Lhasa, was constructed in the mid-seventh century and is considered the holiest site by Tibetans. Its architecture was originally unique, but over time it deteriorated and is now undergoing renovation by the Chinese government. Other notable landmarks in Lhasa include the Lukhang Temple, the Potala Palace (formerly the residence of the Dalai Lama), and the People's Pleasure Park. The two largest monasteries in Tibet, Drepung and Sera, have also been restored.

Lhasa is well-connected by road to Chengdu in Sichuan Province, Qinghai Province, and Xinjiang's Uygur region. The Lhasa airport provides links to other parts of China as well as Nepal. In 2006, a railway line was built, connecting Lhasa with nearby cities in the aforementioned provinces, enhancing accessibility. Tourism has since become a vital part of the local economy, and Lhasa is now recognized as a key destination for cultural and historical tourism in Tibet. Several of its historical sites, including the Potala Palace (designated in 1994) and Jokhang Temple (designated in 2000), are UNESCO World Heritage Sites. As of 2010, the population of Lhasa was 199,159 (Rong 2020).

The flight from Lhasa to Chengdu was scheduled to take 2 h, but due to heavy air traffic at Chengdu's international airport, it was delayed by an additional 50 min. I arrived safely, though nearly 3 hours later than expected. It was around 10:00 p.m., and I was concerned about whether anyone would still be there to pick me up at such a late hour. Fortunately, as I approached the arrival gate, I saw Prof. Deng and his student waiting for me, holding a sign with my name. Prof. Deng greeted me with a warm hug, and we exchanged pleasantries.

It was midnight when I, along with Prof. Deng and Mr. Tony, arrived at the flat where I was going to stay for the next 6 months. The flat, owned by Prof. Deng, had been rented to me at a very nominal rate. It was a two-bedroom apartment, spacious, and well-furnished, featuring a large hall, a dining room, and two bathrooms. The flat also had a piped gas connection and an ample water supply. Located on the fifth floor of the building, there was no lift, so I had to climb the stairs every day. Since we arrived so late, around midnight, I couldn't go out for dinner as all the hotels had already closed. After dropping me off, both Prof. Deng and Tony left, and I was left hungry. On the dining table, there were some instant noodle cups, fruits, and a few bottles of Coca-Cola. I ate the fruit and drank the Coca-Cola. Chinese cuisine generally includes a variety of meats in most dishes, and the noodles were no exception, being non-vegetarian. Pork, beef, rice, and bakery items are the staple foods in China. Sichuan cuisine, in particular, is known for being very spicy and flavorful, and it enjoys worldwide fame. However, distinguishing whether a dish was vegetarian or non-vegetarian proved challenging due to the language barrier, as I couldn't understand Chinese. Regardless, I managed to enjoy a restful sleep that night.

From the next morning, I started self-cooking. China ranks first in the production of vegetables and fruits and therefore, these were easily available, comparatively cheaper, fresh, and plenty. Rice, oil, salt, and some spices were available, which were enough for my daily food need. I had a good kitchen with all utensils including a pressure cooker. Therefore, throughout my stay in Chengdu for 6 months, I enjoyed with self-cooked tasty food and beverages. In the meantime, whenever, I used to go in the field visit, outside Chengdu, it was difficult to me to have vegetarian food.

One day, I went to a restaurant for lunch. The menu was difficult to read as it was written entirely in Chinese. I decided to order boiled rice and spinach curry. When the food arrived, I noticed small brown pieces mixed in the spinach curry. Curious about the ingredients, I asked the waiter to explain what was mixed in the dish, but she was unable to respond clearly. Fortunately, an English couple sitting beside me understood my dilemma and informed me that the brown pieces in my curry were bits of pork. I immediately asked the waiter to replace the dish and bring me a fresh one without any pork. She returned with the same dish, simply removing a few pieces of pork. I paid for the meal but left without eating, feeling disappointed. After this experience, I decided never to visit a restaurant alone again. There were also many occasions when I felt uncomfortable while eating with my colleagues, including Prof. Deng. In Chinese dining culture, people typically eat together using two small wooden sticks. On several occasions, when Mr. Tony ordered vegetarian food for me, he would use the same sticks—those used for eating pork or beef—to pick food from my plate. This made me feel uneasy, as it didn't align with my personal eating preferences.

The lock system in the apartment was automatic. One evening, I went for a walk and accidentally left the key inside the room. When I returned to the apartment, I found I couldn't open the door. By that time, the sun had set, and all the shops were closed. Unfortunately, I didn't have Prof. Deng's phone number with me. I tried to speak to people walking around the apartment complex, but the language barrier made communication impossible. After a while, a young boy noticed my situation, called a locksmith, and helped me get the door open. The entire process took about 4 h.

The days passed with a mix of emotions. I would wake up early each morning and go for a walk, covering four to five kilometers. In the evening, I would take another walk, this time for about three kilometers. My office was on the fifth floor of an old building with no lift. I had three working shifts—morning, after lunch, and after an early dinner—which meant that each day I climbed 250 stairs to reach my office and then descended the same number of stairs. The institute provided me with a desktop computer and high-speed internet, which allowed me to watch a Hindi movie every day. I also learned to digitalize maps using GIS and worked diligently to complete my research project on time. Through collaboration with Prof. Deng and the team members from his lab, I was able to publish several research papers and a book.

I developed a liking for walnuts and peaches, which became my favorite fruits, and they were available in abundance. Among vegetables, peas, taro root, potatoes, eggplant, cauliflower, spinach, and capsicum were my top choices. For dinner, I usually had Chinese steamed bread, which I bought from roadside shops every evening, paired with a self-cooked vegetable curry. For lunch, I typically ate rice with curry. I would buy hot boiled rice from a local restaurant and cook the curry myself. This routine became so familiar that many restaurant waiters and street vendors recognized me. In Chengdu, there were four Indian restaurants, but I only visited them occasionally. One evening, I decided to have dinner at one of these restaurants, which was located near our institute. To my surprise, the owner of the restaurant was Mr. Khanduri, a native of Dehradun, my hometown. As we began talking, I discovered that he was from the same remote region of the Himalayas where I grew up and that he had also graduated from the same college where I had studied.

During the early days of my stay in Chengdu, I often felt lonely and nervous. There were times when I didn't speak a single word for an entire week. The Chinese people, however, were humble and well-behaved, and they took a liking to me, calling me "Indu." Whenever I went to and from the apartment, they would try to engage with me, greeting me with "Nihau" (Hello). I would respond with "Nihau," followed by laughter from both sides, "Ha... ha... ha..." It became a routine. Whenever I walked on the street and interacted with the locals, I would say, "Indu, I am Indu." This continued for the full 6 months of my stay.

During my 6-month stay in Chengdu, I observed that the Chinese people do not follow any particular religion in the traditional sense. Instead, their devotion is to their nation, and they live for the collective purpose of nation-building. Although Buddhism once flourished in China, its followers are now few and mostly reside in the southwestern regions, including Tibet, Sichuan, and Yunnan provinces. Today, China is governed by a single-party system, which is the largest political party in the world.

I also noticed that Chinese scientists are grounded and exceptionally hardworking. Early in my time at the Institute of Mountain Hazards and Environment (IMHE), I witnessed something that left a lasting impression. One day, I saw a senior scientist carrying a broom, a cloth, and a water bucket to his office. Curious, I asked what he was doing, and he explained that he was cleaning his office. He added that every Sunday, he takes time to clean his workspace. This small but significant act reflected the strong sense of duty and commitment that Chinese citizens have toward their responsibilities and the broader goal of nation-building.

Chengdu, the capital city of Sichuan Province, is one of the most beautiful cities in China. Sichuan itself is a basin surrounded by mountains and is famously home to the world-renowned endangered pandas. The Minjiang River flows gracefully through the heart of the city, adding to its charm. Chengdu's airport is one of the largest in China, and the city is well-connected by four major railway stations, each situated at one of the city's four corners. The city's infrastructure is world-class, with wide, well-maintained roads, towering multistoried buildings, lush parks, bustling malls, top-tier hotels, and a wide array of restaurants, all of exceptional quality. Chengdu's cuisine is distinctive—hot, spicy, and delicious—making it a culinary haven. Tourists who visit Chengdu often find it hard to leave, drawn to the city's unique blend of natural beauty, modern infrastructure, and vibrant culture.

The areas surrounding Chengdu are home to many renowned tourist destinations. Among the most famous are the Laughing Buddha, Snow Mountain, Mao Shang, Wenchuan, and Mount Gonga. Additionally, Chongqing City and the Three Gorges Dam are also accessible from Chengdu. During my 6-month stay, I had the opportunity to visit all of these remarkable places, each offering its own unique beauty and cultural significance. These trips added to the richness of my experience in Chengdu, allowing me to explore both the natural and historical wonders of the region.

A research project on climate change and its impact on agriculture in Xichang was sanctioned by CAS, Beijing. Shortly after I joined IMHE, our 20-member team traveled to Xichang for the project. Prof. Deng always participated in every task, no matter how big or small. The field visit lasted for a week. While the journey by road took a full day, the flight was just an hour long. Prof. Deng, along with his student Dr. Song, preferred to travel by air, while the rest of us traveled by a small bus. Once in Xichang, our accommodation was arranged in a five-star hotel. Every morning, we would head to the field for our investigations and return to the hotel in the evening. This routine continued daily until the field investigation was completed.

The Visiting Professor fellowship for foreigners was quite generous, exceeding the salary of local professors. In addition to the fellowship, the other allowances were substantial. Daily allowances for field visits, participation in conferences, and remuneration for giving lectures were more than adequate. Nearly every week, I had the opportunity to give a lecture to Ph.D. students. I also participated in several conferences outside Chengdu, including those in Baoshan, the border area of Myanmar, Chongqing, and Moxi, which is located near the Tibet border. During this period, I also had the chance to visit numerous tourist destinations, making the experience both professionally enriching and personally fulfilling. A variety of local food and beverages are served in Moxi market, dominated by Tibetans (Fig. 8.3).

Prof. Deng Wei was a prominent figure in the academic world in China, well-respected for his contributions to the field. Despite his considerable reputation, he was remarkably humble and down-to-earth. His English language skills were not strong, yet his influence and popularity in China were undeniable. He served as both a Professor and the Director of IMHE for two decades, leaving a lasting impact on the institute. Although I had met Prof. Deng on two previous occasions—in 2012 and 2014—this meeting stood out as particularly memorable and meaningful.

Prof. Deng and I shared a bond that was like that of two brothers, transcending borders and cultures. During my stay in Chengdu, he was a constant source of support, guiding me like a guardian. Every weekend, we would enjoy meals together, and I visited his home several times. He had a son working in a company in Beijing,

Fig. 8.3 A variety of spicy food items in Moxi

and he and his wife lived in a beautiful house away from IMHE. In the evenings, during our walks, he would always help me cross the road. Our friendship remains strong to this day, though we now seldom communicate through email.

Prof. Deng has retired from IMHE and now holds a position as a professor at Chengdu Normal University. His son is married and lives in Beijing, while Prof. Deng and his wife have grown older. I still cherish the memories of the time we spent together in Chengdu. I returned to India in January 2017, which marked our last official meeting concerning our cooperation. However, we met again in 2018 and 2019 in Kunming and Chengdu, respectively, before the outbreak of the COVID-19 pandemic.

In 2010, I was awarded the Visiting Scholar award by The World Academy of Sciences (TWAS), Trieste, Italy, in collaboration with the Chinese Academy of Sciences (CAS), and my workplace was IMHE, Chengdu. During this visit, I worked on livelihood sustainability in the earthquake-affected region of Wenchuan County with Prof. Song Cheng, a Canadian citizen of Chinese origin. The visit was very successful, and upon returning to India, I was again awarded a Visiting Scholar award from CAS. This time, my collaborator was Prof. Deng Wei.

I had previously visited Chengdu in 2007 to participate in an international disaster conference, where I was part of a large and diverse group, including Dr. Tang, Dr. Bao, Dr. Tony, Dr. Luya Feng, Dr. Fu Bin, and Dr. Liu. The second visit in 2014 took me to Mao Shan County along the Minjiang River. Mao Shan is renowned for its beautiful cultural and natural landscapes and is a popular tourist destination in Sichuan. It produces a variety of fruits and vegetables, although the region is highly prone to geo-hydrological disasters. Despite this, the infrastructure in the area is world-class. The Minjiang River is home to several hydropower projects, providing electricity to both rural and urban areas along its valley. Mao Shan is located 160 km from Chengdu, with 60 km of road built on pillars and 70 km passing through tunnels. The area had suffered a devastating earthquake in 2008, with a magnitude of over 8 on the Richter scale, resulting in the loss of more than 200,000 lives and extensive property damage. However, the Sichuan Government rebuilt the affected towns and villages within 2 years, greatly improving infrastructure.

My first visit to China after the COVID-19 pandemic took place in April 2023, when I participated in the Third Mountain Future conference held in Kunming from 16–18 April 2023. The journey was a long one, lasting more than 30 h. The direct flight from Kolkata to Kunming, which typically takes just 2:30 h, had been suspended due to the pandemic. As a result, I had to travel via Kuala Lumpur, which made the trip considerably longer. Upon arriving in China, I encountered a very different situation. There were rigorous checks at Kunming airport that lasted for about an hour. However, the airport staff were extremely supportive and efficient throughout the process. From the airport, the Expo Grand Hotel—where the conference was being held—was located about 30 km away. The organizers arranged for transport from the airport to the hotel, where I was warmly welcomed and provided accommodation. The volunteers at the hotel handed me a registration kit and served meals. Despite having visited China 16 years earlier, I noticed that the Chinese understanding of vegetarian food had not changed much. For them, vegetarian food typically consists of green leaves, groundnut seeds, maize, and fruit juice.

On the first day of the conference, during the inauguration session, I had the opportunity to meet the Tibetan Scholars Delegation. However, not all of them spoke English. When I introduced myself as being from India, they were visibly delighted, and each of them hugged me individually. This experience made me realize that people from around the world generally have a positive perception of India, and they warmly welcome interactions with Indians. Interestingly, despite the recent tensions between India and China, the organizers and attendees continued to show great respect for Indians, and their behavior remained unchanged, reflecting the enduring warmth and goodwill toward our country.

The four-day conference was both exciting and productive. After its conclusion, I returned to the airport at 8:00 am for my 12:30 flight to Bangkok. The process at the airport was quite rigorous, beginning with an electronic declaration, followed by baggage screening, immigration, and a final security check. At immigration, the officer asked me several questions before taking my fingerprints and photo. My trolley bag was screened twice during the security check. Inside the airport, most of the shops were closed, and the overall atmosphere was quite empty, reflecting the severe impact that the Covid-19 pandemic had on China. In December 2023, I traveled to Beijing once again to attend an international conference. It was a five-day visit, which turned out to be both fruitful and enriching.

8.3 Visiting Mount Gonga Region

During my stay at IMHE, Chengdu, I had the opportunity to travel to Mount Gonga (7000 m), located on the Tibet border. The journey began from Chengdu to Moxi city, following the course of the Minjiang River, which took about 12 h. The river valley landscape along the way was breathtaking, with panoramic views that were truly remarkable. The roads were of high quality, and there were restaurants along the route. Tibetan houses are traditional and beautiful (Fig. 8.4).

Tibetan has traditional houses, made of stone and wood. They are designed in such a way that the climate cannot influence them much. At every evening, the women dance in the community common place with their traditional dresses and they attract tourists (Fig. 8.5).

There were around 50 participants from various countries, traveling in two luxury buses to attend an international conference in Moxi. The event was scheduled for 4 days. Moxi is a small hill city with strong Tibetan cultural influences. The local people, mostly of Tibetan descent, practice farming and animal husbandry. The climate is temperate, though mild in the summer. Rice and maize are the main crops grown, while yaks, goats, sheep, and lambs are raised by the locals.

The first 2 days of the conference were dedicated to academic presentations. However, I found it difficult to eat during the lunch and dinner buffets, as many of the food items were mixed with meat or boiled, and few vegetarian options were available. Tibetans commonly eat yak meat and even insects, though they do serve hotpots with vegetables for tourists who are vegetarians. Due to the cold climate, many locals have reddish faces with black spots.

Fig. 8.4 Modern Tibetan house with temple in the ground floor

(a) (b)

Fig. 8.5 (**a**) Traditional Tibetan village and (**b**) Tibetan women dancing in a common place at evening

Fig. 8.6 Mount Gonga below which is the glacier and barren and stony land

On the third day, we embarked on a field trip to Mount Gonga. We started early in the morning, traveling by ropeway up to the halfway point. After an hour-long journey from Moxi, we arrived at a location where the Department of Botany at IMHE has a laboratory dedicated to plant pathology. The organizers offered us beverages, and after a short break, we began our trek toward Mount Gonga. The trek was about 5 km through a perennial stream and thick deodar forests. The terrain was rugged, with steep slopes that made the journey challenging.

After a strenuous hike, we reached the other side of Mount Gonga at an altitude of 5000 m. Beneath the mountain, a massive glacier lay before us, and we stood facing its towering ice. The altitude of Mount Gonga is 7000 m, making it the highest peak in the region, which borders Tibet (Fig. 8.6). The weather was frigid, and we were all bundled up in warm jackets and other layers to protect ourselves from the cold. It was an unforgettable experience, surrounded by the grandeur of nature at such an impressive height.

The site was truly magnificent, thoughtfully developed for visitors. The trails were made of sturdy wooden planks, and every 100 m along the path, a well-maintained washroom facility was available, equipped with all necessary amenities. Workers were actively collecting any trash left behind by tourists, ensuring the area remained clean. The landscape, with small trees, rocks, and the gentle slopes, provided a breathtaking panoramic view that captivated everyone. Most of the group spent their time taking photographs to capture the beauty of the place.

Two small shops were set up to serve tea and snacks, adding a pleasant touch to the experience. After spending 2 h enjoying the surroundings, we began our trek back to Moxi. On the way, we stopped by hot spring ponds where we took a relaxing bath. On the final day, we stayed in Moxi again, before heading back to Chengdu the following morning. The entire trip to Moxi and Mount Gonga was a wonderful experience, filled with natural beauty, peaceful moments, and memorable adventures.

8.4 From the Summit of Long Meng

Time is indeed a continuous, ever-moving force, never pausing or signaling its arrival or departure. It flows undisturbed, unaffected by our actions, yet it carries immense power. Time does not judge or show favoritism; it simply moves, offering opportunities and consequences in equal measure. We are fortunate when we are able to recognize and make the most of the opportunities that time grants us. I feel particularly blessed for having had the chance to experience some remarkable moments, one of which was my visit to Maoshan County, Sichuan Province, China. It was an opportunity that not only broadened my horizons but also provided invaluable insights into both the region's rich cultural landscape and its unique environmental challenges.

In August 2014, a research team led by Prof. Deng Wei, then Director of IMHE, CAS, Chengdu, embarked on a field survey in Maoshan County, Sichuan Province. The team, which included Dr. Song, Dr. Luya, Dr. Fubin, and myself, spent a week surveying households in a local village. On the final day, we continued our usual routine of gathering data in the morning before having a typical spicy Sichuan lunch. Despite my high tolerance for Indian spices, navigating the local Chinese cuisine was a challenge, especially as a vegetarian. I often found myself surviving on noodles, a simple yet reliable meal.

With only 7 h of daylight left, I suggested a visit to the local apple orchard to capture a few photos. However, the suggestion quickly morphed into a more ambitious plan. The Chinese scholars enthusiastically proposed a trip to Maoshan's highland alpine meadows, located 60 km from our current location. We set off for Wenchuan city and then toward the alpine pastureland. The road split just after

Fig. 8.7 Apple orchards and horses being reared in pastureland. Photo By author

crossing the Minjiang River, following the riverbank for a while before curving upward. The road was in excellent condition, allowing us to drive at a good speed.

As we ascended, the landscape became increasingly breathtaking. The view was dotted with beautiful mountain settlements, each surrounded by orchards of temperate fruits like apples, peaches, plums, apricots, walnuts, and almonds (Fig. 8.7). Maoshan County's climate is ideal for fruit cultivation, and fruit farming plays a crucial role in the region's economy, which is often referred to as the "Tree Economy." Due to the high yields of these fruit crops, many of the local farmers have become quite prosperous.

After traveling 40 km, we reached the point where human settlements faded away, replaced by dense temperate forests. Continuing further, we entered the alpine grasslands. This region, known as "Long Meng," is one of China's ten national natural reserve areas. As we ascended another 10 km, the area began to unfold its full splendor. Along the way, we spotted various faunal species, with horses standing out as a major attraction. The journey was both enriching and awe-inspiring, giving us a glimpse into the natural beauty and ecological importance of the Maoshan highlands.

China's infrastructure is often regarded as one of the best in the world, and the pastureland we visited was no exception. The government had constructed ropeways that reached two different peaks, enhancing the accessibility of the area. Complementing the natural beauty were well-maintained wooden bungalows, classic toilets, and aesthetically pleasing signboards, all adding to the overall charm of the location. However, the rainy and cloudy day made the pedestrian paths slippery, which, though picturesque, posed a safety risk as we navigated along the wooden pathways. The alpine grasslands, teeming with colorful flowers and medicinal plants, only heightened the appeal of the landscape. The air was fresh and fragrant, evoking a deep sense of connection to nature. The entire scene seemed to transport me mentally and emotionally to the Himalayas, evoking memories of my past experiences there. In that moment, I felt a profound spiritual connection to the Himalayas, which, in my heart, is an embodiment of Lord Shiva. It was an overwhelmingly emotional and spiritual experience, one that resonated deeply within me.

Fig. 8.8 White mountain and alpine pastureland covered by flowers and medicinal plants

After walking about a kilometer, we came upon a dry highland lake located at an altitude of 3000 m (Fig. 8.8). The lake, named Bai Long, or "Dragon" in Mandarin, was once a perennial body of water. The name is significant in Chinese culture, as the dragon is revered as a folk deity and worshiped by locals. Tragically, the lake was severely affected by the 2008 Wenchuan earthquake. The powerful tremors caused a fault to form in the river, splitting it and draining the water from the lake. The once full lake was now a dry basin, with only a high-volume stream flowing through the middle. Chinese geologists were actively working to restore the lake, but the damage from the earthquake remained a stark reminder of the region's vulnerability to natural disasters.

We enthusiastically explored the surroundings of the lake, completely in tune with nature's rhythm. Alongside us, a few domestic tourists were enjoying the moment, laughing, dancing, and singing songs with their families, adding a light hearted energy to the serene landscape. After spending an hour in the pastureland, the clouds began to clear, revealing the majestic Long Meng mountain range in the distance. Some clouds still lingered, gently kissing the White Mountain. The sight was breathtaking—the mountain range appeared white but was barren, surprisingly without snow. The scene was so striking that our team stood mesmerized. I, too, was deeply moved; time itself seemed to pause.

What had seemed like the highest peak from our vantage point now made us realize that even though we were at a high altitude, we were still surrounded by taller mountains. The sun had started to set, and darkness began to unfold, urging the light to retreat. The woods around us grew dark and deep, adding to the quiet majesty of the moment. We soaked in the peace, lost in our thoughts, choosing not to speak. As I slowly made my way back, I found myself silently reciting the famous lines from Robert Frost's poem: "*Miles to go before I sleep*..." My heart was full, and my mind was brimming with memories that felt eternal. It was a miraculous day—one that will remain etched in my heart forever.

8.5 Conclusions

My observations about China and the Chinese people are rooted in the profound sense of humanity that transcends borders and cultural differences. Humanity, in its purest form, is always greater than any culture or geographic boundary. This memoir is a reflection of my ten visits to Tibet and China, where I had the opportunity to engage in meaningful conversations and exchanges with the local people. Through these interactions, I came to understand that at the core of every culture, there is a shared humanity that binds us all, a bond that is not confined by national borders or cultural identities.

References

Cao G, Yeh ET, Liu Y, Zhou H (2013) Tibetan herders' perceptions of grassland degradation and climate change in the eastern Tibetan plateau. Reg Environ Chang 13(1):123–132. https://doi.org/10.1007/s10113-012-0321-2

Chen F, Chen J, Holmes J, Boomer I, Austin P, Gates JB, Yu Z (2010) Moisture changes over the last millennium in the northeastern Tibetan plateau: a review of proxy records. Quat Sci Rev 29(25-26):3312–3321. https://doi.org/10.1016/j.quascirev.2010.08.007

Fischer AM (2014) The disempowered development of Tibet in China: a study in the economics of marginalization. Lexington Books, Lanham

Foggin JM (2008) Depopulating the Tibetan grasslands: national policies and perspectives for the future of Tibetan herders in Qinghai Province, China. Mountain Res Dev 28(1):26–31. https://doi.org/10.1659/mrd.0972

Goldstein MC (1991) A history of modern Tibet, 1913–1951, the demise of the lamaist state. University of California Press, Oakland, p 639

Harris RB (2010) Rangeland degradation on the Qinghai-Tibetan plateau: a review of the evidence of its magnitude and causes. J Arid Environ 74(1):1–12

Hillman B (2003) Paradise under construction: minority tourism and development in Tibet. J Contemp China 12(37):347–364

Kang S, Xu Y, You Q, Flügel WA, Pepin N, Yao T (2010) Review of climate and cryospheric change in the Tibetan plateau. Environ Res Lett 5(1):015101. https://doi.org/10.1088/1748-9326/5/1/015101

Næss MW (2013) Climate change, risk management and the end of nomadic pastoralism. Int J Sustain Dev World Ecol 20(2):123–133. https://doi.org/10.1080/13504509.2013.779615

Nyima Y, Williams N (2022) Tourism and cultural heritage in Tibet: impacts and community perspectives. Tour Geogr 24(1-2):148–167

Qiu J (2016) Trouble in Tibet: farmers and herders are struggling to cope with environmental and social changes on the world's highest plateau. Nature 529(7585):142–145. https://doi.org/10.1038/529142a

Rong M (2020) Population and society in Tibet. Hong Kong University Press, Pokfulam, p 350

Wang S, Zhou D (2020) Economic transformation and infrastructure development in Tibet: patterns, progress, and implications. China Econ Rev 61:101456. https://doi.org/10.1016/j.chieco.2020.101456

Wang B, Bao Q, Hoskins B, Wu G, Liu Y, Yang Y, He J (2008) Tibetan plateau warming and precipitation changes in East Asia. Geophys Res Lett 35(14):L14702. https://doi.org/10.1029/2008GL034330

Wu N, Yin G, Sun X (2015) Agropastoral transitions on the Tibetan plateau: a review of recent changes and future prospects. Pastoralism 5(1):1–12

Yang Q, Zheng D (2004) Tibetan geography. China Intercontinental Press, Beijing, pp 30–31

Yao T, Thompson LG, Mosbrugger V, Zhang F, Ma Y, Luo T, Joswiak D (2012) Third pole environment (TPE). Environ Dev 3:52–64. https://doi.org/10.1016/j.envdev.2012.04.002

Yeh ET, Henderson M (2008) Interpreting urbanization in Tibet: administrative scales and discourses of modernity. J Contemp China 17(56):131–155

Zhao W, Ritchie BW (2007) Tourism and poverty alleviation in Tibet: a political economy perspective. Tour Manag 28(5):1283–1293

Chapter 9
Conclusions

Contents

Abstract In this section, a summary of all chapters is presented. The cultures, economies, and environments that I studied and observed during my extensive visit across the entire Himalayan region, including the Trans-Himalaya, are illustrated here. Each cultural and natural section of the Himalaya is unique, with distinct characteristics that vary from one region to another. The Trans-Himalaya is a barren, cold desert, while the main Himalayan region is rich in natural resources. The Himalaya is home to diverse religious traditions, with Hinduism and Buddhism being the most prominent. Often referred to as the "Water Tower of Asia" and the "Third Pole," the Himalaya holds immense economic significance. The world's three largest river systems originate and flow through this mountain range.

Keywords Ladakh · Kedarkhand · Manaskhand, Kali Gandaki · Bhutan · Tibet

The Himalaya is one among the mountain ranges in the world that holds both natural and cultural significance. On one hand, it is the source of three major river systems—the Sindhu (Indus), the Ganges, and the Brahmaputra—along with their numerous tributaries (Ives et al. 1997). On the other hand, the Himalaya is believed to be the abode of thousands of folk deities. It is considered the embodiment of Lord Shiva and Shakti, the revered folk deities of Hindus. All natural entities—land, water, and forests—are regarded as sacred and worshiped on various occasions (Bernbaum 1997). These river systems not only provide water but also sustain the livelihoods of over 500 million people (Singh 2008). The Himalaya consists of the Lesser Himalaya, Middle Himalaya, Greater Himalaya, and Trans-Himalaya in a vertical classification. Horizontally, it is divided into three regions: the Western

V. P. Sati, *Exploring the Mystical Himalaya*, Springer Geography,
https://doi.org/10.1007/978-3-032-14275-7_9

Himalaya, the Central Himalaya, and the Eastern Himalaya (Valdiya 2015). The majestic mountain peaks, deep river valleys, and vast wilderness make the Himalaya a breathtaking landscape. However, reaching these regions requires arduous trekking, demanding both physical endurance and enthusiasm. Journeying to the Himalaya is a long and relentless pursuit, requiring unwavering determination and resilience. It may take years to truly explore its vastness. Before embarking on a trek to the Himalaya, one must understand its towering heights, diverse landforms, solitude, challenges, spiritual essence, and abundant natural resources.

The Himalaya is manifested through its diverse landscapes, rivers and streams, snow-capped mountain peaks, and various forests. Its spectacular landforms, the rhythmic flow of rivers and streams shaping different topographies—cascades, waterfalls, rapids, and meanders—create a breathtaking natural spectacle (Ives 2004). The towering mountain peaks change their colors and forms with time, while the panoramic and diverse forest landscapes create multiple forest layers, resonating with the chirping of birds and the calls of wildlife (Singh 2015; Negi 2002). Additionally, the Himalaya serves as a *taposthali* (a place of penance) for sages and saints, a sacred haven for pilgrims, and a life-sustaining region for billions of people.

The treatise *Exploring the Mystical Himalaya: Cultures, Economies, and Environment* illustrates the natural, cultural, economic, and environmental significance of the Himalaya. It covers the region from the Pamir Knot in the northwest to the Arakan Yoma Mountains in the southeast. The book consists of a total of nine chapters, including the introduction and conclusion. The core chapters focus on Uttarakhand, Ladakh, Nepal, Bhutan, and Tibet. Uttarakhand is covered in three chapters, while each of the other regions has one dedicated chapter. All chapters are systematically summarized and presented.

A brief description of each chapter is presented in the first chapter, i.e., the introduction. It provides an overview of the nature, culture, economy, and environment across different parts of the Himalaya. The Himalaya is home to diverse cultures, societies, rituals, and customs, which vary from one region to another. The people of the Himalayan region belong to multiple religions and castes, each with its own distinct fairs, festivals, and traditions. Similarly, economic activities, development patterns, environmental conditions, and landscapes also differ across the Himalaya. The economy of the region primarily depends on traditional agriculture and livestock farming. However, the output from these activities is relatively low, resulting in widespread poverty, with a significant portion of the population living below the poverty line. The introduction chapter provides a comprehensive description of these aspects.

Hari Ki Dun, meaning the "Valley of God," is located in the high Himalaya of western Uttarakhand. The Tons River, a major tributary of the Yamuna River, originates from Hari Ki Dun and flows through the Jaunsar, Bawar, and Rawain areas in Uttarkashi district before joining the Yamuna River at Kalsi. This region forms one of the sub-cultural realms of the Uttarakhand Himalaya, characterized by its unique and diverse cultural heritage. The customs, rituals, language, food habits, fairs, festivals, and attire of this region are rich and distinct from other parts of Uttarakhand. Within the Tons Valley, there are two sub-cultural realms—Jaunsar-Bawar and

Rawain. The people of this region believe they are descendants of the Pandavas and Kauravas, and as a result, polyandry and polygyny are still practiced in some areas of the Tons River basin. The region is home to several folk deities, with temples dedicated to Mahasu, Karna, Duryodhana, Pokhu, and Someshwar Mahadev located along the Tons River basin. Another significant religious site is the Lakhamandal Shiva Temple, situated in the Yamuna River valley. The local population follows Hinduism, and settlements are sparsely distributed. Agriculture remains the primary economic activity, with rich agrobiodiversity. The major crops grown in the highland areas include apples, potatoes, rajma (kidney beans), and millets. However, crop production and productivity remain low, resulting in widespread poverty, with a large portion of the population living below the poverty line. Infrastructural facilities, including transportation and communication, are underdeveloped, especially in the high-altitude areas, despite the region's breathtaking landscapes. Natural and cultural tourism have significant potential for development, provided that adequate infrastructure is put in place.

Kedarkhand is the ancient name of the Garhwal Himalaya, as mentioned in the rich cultural texts of Hinduism. The name Kedarkhand is derived from the famous highland pilgrimage site Kedarnath, located in the Rudraprayag district on the banks of the Mandakini River. This region is renowned for its sacred sites, including the Panch Badris, Panch Kedars, and Panch Prayags. It is also the source of the Ganga River and its many tributaries. Kedarkhand is celebrated for its cultural heritage and temples, with numerous fairs, festivals, customs, rituals, and cultural processions that are both unique and diverse. One of the most famous cultural processions is the Rajjat Jatar, the longest Jatar (pilgrimage procession) of Kedarkhand. Kedarkhand is well known for river valley and highland pilgrimages, where cultural tourism thrives. The region has several prominent cultural routes, including the routes from Haridwar to Gangotri and Yamunotri, Haridwar to Badrinath and Kedarnath, Karnprayag to Homekund, and Haridwar to Pauri and Srinagar. These routes are home to numerous sacred temples dedicated to Shiva, Shakti, and Vishnu, along with several significant Shakti Peeths. In addition to its cultural and spiritual significance, Kedarkhand is rich in biodiversity. It supports a variety of crop cultivars and landraces, with traditional agricultural practices still prevalent in the region (Nautiyal et al. 2003). The forests and natural landscapes are three-dimensional and spectacular, adding to the region's breathtaking beauty and ecological importance.

Manaskhand is the second important region of the Uttarakhand Himalaya, known today as the Kumaon Himalaya. It comprises Western Kumaon and Eastern (Kali) Kumaon. Western Kumaon shares its border with the Garhwal Himalaya and is famous for its spiritual and historical sites such as Jageshwar, Bageshwar, and the Sun Temples. The region is also known for its rivers and highland lakes, which serve as major tourist destinations. Popular hill stations like Nainital, Almora, and Ranikhet, located at middle altitudes, attract visitors throughout the year. The region has a rich cultural heritage, with diverse customs, rituals, fairs, and festivals celebrated across seasons. The people of Western Kumaon follow Hinduism and belong to various castes and communities. Agriculture, along with livestock farming, is the primary economic activity, yielding better output compared to the Garhwal

Himalaya. Major rivers flowing through the region include the Gomati, Saryu, Ramganga (West), and Kosi. A part of Nainital and Udham Singh Nagar districts consists of flat land, which is highly suitable for agricultural farming. Eastern Kumaon, known as Kali Kumaon, shares its borders with Tibet in the north and Nepal in the east. The Kali River (also known as Choka and Sharda) forms the natural border between Kali Kumaon and Nepal, running from north to south. Due to this geographical proximity, the border communities share cultural and social ties, including intermarriages and similar food traditions. The major tributaries of the Kali River are Ramganga (East), Gori Ganga, and Dhauli Ganga (East). The districts of Pithoragarh, Champawat, and the eastern part of Udham Singh Nagar make up Kali Kumaon. This region is home to three significant valleys—Johar (Gori Ganga), Darma (Dhauli Ganga East), and Vyas (Kali River)—each holding both natural and cultural significance. The highland pastures (Bugyals) serve as grazing grounds for livestock farming, with locals primarily rearing goats, sheep, lambs, and horses, following seasonal migration. In the river valleys, people practice traditional agriculture, cultivating cereal crops, including millets, to sustain their livelihoods.

Ladakh, also known as Leh Ladakh, is a Trans-Himalayan region located in northern India. It shares borders with Tibet in the east, Jammu and Kashmir in the west, Himachal Pradesh in the south, and the northern part is adjacent to Pakistan-occupied Kashmir. It is an integral part of the Republic of India. The landscape of Ladakh is breathtaking, characterized by wet river valleys, dry and barren middle altitudes, and snow-covered highlands throughout the year. Ladakh is a high-altitude plateau, with elevations starting from 3400 m above sea level. Forests are sparsely distributed, mainly found near water bodies, forming oases in the arid landscape. A large part of Ladakh is desert-like, with harsh climatic conditions. Winters are extremely cold, with temperatures dropping below freezing, while summers can be intensely hot. Due to the arid climate, sand dunes are formed in the Nubra Valley. The Sindhu (Indus) River dominates the region's drainage system, with numerous glacier-fed tributaries. Considered holy, the Sindhu River is worshiped by the people of Ladakh on various occasions. It is also a major water source for both India and Pakistan. Other significant rivers include the Zanskar and Shyok Rivers. The Zanskar River merges with the Sindhu River near Leh, while the Shyok River flows directly into Pakistan. The Nubra River is a tributary of the Shyok River. Ladakh has an ancient history and has been ruled by various dynasties, including Muslim rulers from different directions. Consequently, around 50% of the population follows Islam. Ladakh has also been significantly influenced by Tibetan culture, and Tibetans once ruled the region. Approximately 50% of the population practices Buddhism, with ten famous monasteries serving as major places of worship and spiritual learning. Historically, Ladakh was a key trade center for Central Asia. However, following the Chinese aggression of 1962, all trade relations with Tibet and Central Asia ceased. The economy of Ladakh is primarily based on livestock rearing, with yak, cattle, goats, sheep, lambs, and horses being the main livestock in the highlands. In the river valleys, people practice agriculture and horticulture though low productivity remains a challenge. Apart from farming, locals are also engaged in trade and tourism, with tourism emerging as a vital sector for economic development. Ladakh has numerous tourist attractions, drawing thousands of visitors throughout the year. Some of the major destinations include Pangong Lake, Nubra Valley, Kargil, and Leh. The culture of Ladakh is vibrant, with numerous fairs and festivals celebrated year-round. Leh, the capital city, is well-connected by air and road, with flights from various parts of India operating to Leh. Given its strategic location bordering Pakistan and China, Ladakh holds immense geopolitical significance for India.

Nepal is a tiny, landlocked country and an integral part of the Himalaya. It is known as the Nepal Himalaya or the Central Himalaya. Most of the country is mountainous and snow-capped. Everest and Purnagiri are two mountain peaks located in Nepal. It is drained by many glacier-fed rivers of the Ganges system, creating panoramic river valleys. A large part of Nepal is covered by highly valuable forests, comprising subtropical and temperate forests, as well as alpine grasslands. The climate varies from subtropical to temperate, cold, and frigid cold. Nepal is a Hindu country, but Buddhists also form a significant population. The country has a rich cultural heritage, blending Hindu and Buddhist traditions. The Pashupatinath Temple in Kathmandu is the seat of Lord Shiva, one of the Jyotirlingas, located on the banks of the Bagmati River, a holy river of Nepal. This temple is used for cremations. Another cultural destination is Muktinath, located in the upper Mustang region. The temple is devoted to Lord Vishnu and is situated on the banks of the Kali Gandaki River, another holy river of Nepal, known for the Black Shaligram stones. Both the Bagmati and Kali Gandaki rivers irrigate vast agricultural lands and are culturally significant. A journey from Kathmandu, the land of Lord Shiva, to Muktinath, the land of Lord Vishnu, is unique and can be traversed by air, road, or trekking. The Hindus in Nepal follow both Shaivism and Vaishnavism, as the country is home to many temples dedicated to Lord Shiva and Lord Vishnu. The economy of Nepal is primitive, with people practicing traditional agriculture to sustain their livelihoods. Nepal has rich agrobiodiversity, with the cultivation of food grains, pulses, oilseeds, fruits, vegetables, and spices. Livestock farming is also prevalent, providing a base for sustainable dairy farming. Tourism is another livelihood option. The Himalayan rivers are perennial, providing water for irrigation, hydropower projects, and drinking water. Nepal is home to thousands of trails, passes, rivers, streams, rivulets, waterfalls, gorges, mountain peaks, cultures, customs, and rituals. The Himalaya protect the country, which is why Nepal is the only country in South Asia that has remained independent at all times.

Bhutan, a tiny and landlocked country, has a rich cultural heritage. Influenced by Buddhism, Bhutan is the only country in the world where all citizens follow the path of Lord Buddha, with a small population of Nepali people following Hinduism. There are many monasteries and a few temples representing both Hinduism and Buddhism. Bhutan celebrates fairs and festivals throughout the year. The country follows a monarchy, ruled by the King. It is primarily mountainous, with most of it covered by the Greater Himalaya. As a result, it is also called the Bhutan Himalaya. Some areas consist of river valleys and Duar plains. The landscape is spectacular, featuring thousands of landforms, including mountain peaks, passes, trails,

"U"-shaped valleys, "V"-shaped valleys, gorges, waterfalls, rivulets, river terraces, and flood plains. Faunal and floral diversity is rich, with vegetation ranging from subtropical to temperate and alpine grasslands. The economy is primarily based on agriculture, livestock farming, and tourism. A number of crop varieties are grown; however, crop production and productivity are low and insufficient. Crops are grown using traditional methods, with cereals being the dominant crops. Among fruits, apple, pear, peach, apricot, walnut, and plum are prominent, while citrus fruits are grown in the valleys. Yak is the primary livestock reared in the higher altitudes, with cattle, goats, sheep, lambs, and horses also raised. Tourists visit throughout the year, contributing to the economy and development of Bhutan. There are many trails and trekking routes in the Himalaya, where tourists enjoy trekking and mountaineering. For citizens of India and Nepal, entry to Bhutan is free. Bhutan's infrastructure is limited; roads are the main means of transportation, though many remote areas are not connected by road. Railways are limited to connecting the plain areas. There is only one airport, Paro, which connects Bhutan with South Asian countries. A large portion of the population is poor, with many living below the poverty line. People live harmoniously with nature, enjoying whatever resources are available, which is why Bhutan has the highest happiness index in the world. Thimphu, the capital, is a multicultural city, known for its diversity in food, beverages, languages, and clothing.

This book also includes the study of Tibet, the highest and driest plateau in the world. It is part of the Trans-Himalaya and is a cold desert. Tibet was an independent country before 1956, after which China invaded and ruled it. Today, Tibet is an autonomous province of China. The landforms of Tibet are spectacular, though many areas are barren and stony, creating plateaus and river valleys. The top land remains snow-capped, feeding numerous rivers in the region. The altitude ranges above 3400 m, with some areas exceeding 6000 m. Climatic conditions are harsh, with frigid winters and hot summers. Forests are sparsely distributed, mainly found along riverbanks, and are temperate forests, above which alpine grasslands are found. Lhasa, the capital city, is located above 4000 m. The history of Tibet is ancient and has been ruled by many rulers. The people belong to the Tibeto-Mongol race. Since they live in a high-altitude region, their economic activities, culture, food habits, and clothing are quite distinct. They eat meat and drink wine. The people grow a few crops, mainly maize and potatoes, along with some vegetables, fruits, and medicinal plants, and also rear livestock. Yak is the principal animal, used for meat, milk, and wool. Other animals include goats, sheep, lambs, and horses. In addition to providing meat, milk, and wool, these animals are also used for transportation. Trade in jewels and gems is another occupation. Tourism plays an important role in Tibet's economy and development. The economy of Tibet is primitive, and the development of infrastructure is minimal. The climate, landscape, and terrain further hinder economic development. However, Tibet has thousands of rivers, streams, rivulets, waterfalls, gorges, trails, passes, mountain peaks, and trekking routes. These natural features hold great potential for tourism development. The population is sparsely distributed, and the population density is low. People follow Buddhism, and their cultural heritage is rich. There are many world-class monasteries, and the capital city of Lhasa preserves its culture through these monasteries. Tibetans share a deep affinity with Indians due to Buddhism and Lord Buddha. Historically, Tibet had trade and cultural relationships with India. However, with the advent of China in Tibetan territory and its subsequent occupation, the trade relationship ceased, and Tibetans are now struggling to preserve their identity. As a minority population, their cultural identity is gradually deteriorating.

The description above of the Himalaya and its different regions illustrates the incredible diversity found in this mountain range. The Himalaya is home to diverse climates, landforms, economies, environments, populations, cultures, and religions. It boasts the world's highest mountain peak, Mount Everest, and the world's largest river systems—the Sindhu, Ganga, and Brahmaputra. The region also features the world's largest river valley plains—the Ganges and Brahmaputra—and serves as a "water tower" and the "third pole." The Himalaya is the birthplace of Hinduism and Buddhism, a biodiversity hotspot, and a place rich in cultural heritage and diversity. It support the livelihoods of more than 600 million people, regulate climate, preserve natural resources, and safeguard cultural traditions. The Himalaya is one of the most peaceful regions, where culture thrives, and human civilization reaches its highest potential. On the other hand, the Himalaya exemplify unity in diversity. The people live in harmony with nature, respecting and worshiping all natural elements—land, water, air, forests, food grains, mountains, and animals. It is a land where gods and goddesses manifest in human form and dance with nature. The Himalaya is dotted with thousands of trails, and this book seeks to explore the mystical beauty of the region.

References

Bernbaum E (1997) Sacred mountains of the world. University of California Press, Oakland

Ives JD (2004) Himalayan perceptions: environmental change and the Well-being of mountain peoples. Routledge, London

Ives JD, Messerli B, Spiess E (1997) Mountains of the world: a global priority. The Parthenon Publishing Group, Nashville

Nautiyal S, Rao KS, Maikhuri RK, Saxena KG (2003) Traditional knowledge related to agroforestry and social forestry practices in Uttaranchal Himalaya. Indian J Tradit Knowl 2(3):300–310

Negi SS (2002) Himalayan forests and ecological development. Indus Publishing, New Delhi

Singh S (2008) Environmental geography. Prayag Pustak Bhawan, Allahabad

Singh RB (ed) (2015) Climate change and biodiversity: proceedings of IGU Rohtak conference. Springer, Cham

Valdiya KS (2015) The making of India: geodynamic evolution. Springer, Cham

Index

V. P. Sati, *Exploring the Mystical Himalaya*, Springer Geography,
https://doi.org/10.1007/978-3-032-14275-7

MIX
Papier aus verantwortungsvollen Quellen
Paper from responsible sources
FSC® C105338
FSC
www.fsc.org

If you have any concerns about our products,
you can contact us on
ProductSafety@springernature.com

In case Publisher is established outside the EU,
the EU authorized representative is:
Springer Nature Customer Service Center GmbH
Europaplatz 3, 69115 Heidelberg, Germany

Printed by Libri Plureos GmbH
in Hamburg, Germany